Biotechnology and Bioengineering

Edited by Eduardo Jacob-Lopes and Leila Queiroz Zepka

Published in London, United Kingdom

IntechOpen

Supporting open minds since 2005

Biotechnology and Bioengineering
http://dx.doi.org/10.5772/intechopen.77540
Edited by Eduardo Jacob-Lopes and Leila Queiroz Zepka

Contributors
Kenneth Yongabi Anchang, Henrik Toft Simonsen, Betty Revon Liu, Yue-Wern Huang, Han-Jung Lee, Kwang-Il Lee, Kwang-Won Lee, Ju-Woong Jang, Memory Tekere, Noemi Zaritzky, Alejandro Caravelli, Juan Carlos Alzate Marin, Nelson Enrique Arenas Suarez, Luz Mary Salazar, Selvan Ravindran, Pooja Singh, Sanjay Nene, Nuton Mhetras, Vinay Rale, Anuradha Vaidya, Yu (Sherry) Xie, Martin W. King, Harshini Ramakrishna, Jiyang Chen, Fan Zhao, Monica Deshpande, Ting He, Fan Zhang, Katalin Bélafi-Bakó, Peter Bakonyi, Cleide Barbieri De Souza, Gabriel Rodrigues Silva, Eduardo Jacob-Lopes, Leila Queiroz Zepka

First published in London, United Kingdom, 2019 by IntechOpen
IntechOpen is the global imprint of INTECHOPEN LIMITED, registered in England and Wales, registration number: 11086078, 7th floor, 10 Lower Thames Street, London, EC3R 6AF, United Kingdom
Printed in Croatia

British Library Cataloguing-in-Publication Data
A catalogue record for this book is available from the British Library

Additional hard and PDF copies can be obtained from orders@intechopen.com

Biotechnology and Bioengineering
Edited by Eduardo Jacob-Lopes and Leila Queiroz Zepka
p. cm.
Print ISBN 978-1-78984-039-1
Online ISBN 978-1-78984-040-7
eBook (PDF) ISBN 978-1-83962-661-6

For EU product safety concerns:
IN TECH d.o.o., Prolaz Marije Krucifikse Kozulić 3, 51000 Rijeka, Croatia, info@intechopen.com or visit our website at intechopen.com.

Meet the editors

Prof. Eduardo Jacob-Lopes is currently an associate professor at the Department of Food Science and Technology, Federal University of Santa Maria. He has more than 18 years of teaching and research experience, and he is a technical and scientific consultant for several companies, agencies, and scientific journals. His 500 publications/communications include 8 books, 30 book chapters, 85 original research papers, and 360 research communications in international and national conferences, and he has registered 15 patents. His research interests include biotechnology and bioengineering with emphasis on microalgal biotechnology.

Prof. Leila Queiroz Zepka is currently an associate professor at the Department of Food Science and Technology, Federal University of Santa Maria. She has more than 15 years of teaching and research experience, and she is a technical and scientific consultant for several companies, agencies, and scientific journals. Her 420 publications/communications include 8 books, 30 book chapters, 70 original research papers, and 300 research communications in international and national conferences, and she has registered 10 patents. Her research interests include microalgal biotechnology with emphasis on biomolecules and natural pigments.

Contents

Preface

The book *Biotechnology and Bioengineering* shows how biotechnology and bioengineering can be used to solve the demands of modern society. In these eleven chapters, the book explores several important topics of biotechnology and bioengineering such as bioinstrumentation, biomaterials and nanotechnology, cell and tissue engineering and design, and application of bioreactors. The chapters are intended to help provide a deeper and wider understanding of the promises and challenges for biotechnology and bioengineering. This book is primarily dedicated to scientists, academicians, industrial representatives, and innovative technology representatives, and also to any nonspecialist reader willing to learn about the dynamic field of biotechnology and bioengineering.

Eduardo Jacob-Lopes and Leila Queiroz Zepka
Food Science and Technology Department,
Federal University of Santa Maria,
Santa Maria, RS, Brazil

Chapter 1

Introductory Chapter: Biotechnology and Bioengineering

Rosangela Rodrigues Dias, Leila Queiroz Zepka and Eduardo Jacob-Lopes

1. A general overview on biotechnology and bioengineering

Biotechnology and bioengineering can be defined as "the integration of natural sciences and engineering sciences in order to achieve the application of organisms, cells, parts thereof and molecular analogues for products and services" [1]. Although these areas overlap, depending upon the use of techniques and their applications, both have peculiar characteristics. While the focus of bioengineering is the implementation of engineering principles and design concepts in biology, biotechnology is more focused on the natural sciences [2].

These fields, today considered priority, are the fruit of strategies from various areas that aim to benefit humanity and its environment. Biotechnology and bioengineering, however, despite growing attention in recent decades, are not a new science. Humans have been developing them since the earliest beginnings, mainly in food production. Some ancestral examples include the preparation of fermented beverages from cereals in Babylon and Egypt (8000 to 6000 years BC); the production of bread, using ferment in Egypt (4000 years BC); and wine production in Greece (2000 years BC). Historically, the use of these traditional techniques in this period of history is called discoveries, and not development, once the underlying scientific principles were not understood [3, 4].

Indeed today, the biotechnology and bioengineering based on scientific progress find applications in several areas, including agriculture, livestock, human health, preservation of the environment, and manufacturing industry [5]. This wide applicability was only possible due to the combination of several fields of knowledge that include biochemistry, physiology, genetics, microbiology, virology, botany, zoology, ecology, computer science, and chemical engineering.

Therefore, this is a field of work typically multidisciplinary, which makes the effective collaboration and integration of professionals from different areas of knowledge absolutely indispensable so that all potential of biotechnology and bioengineering can be exploited. The interface between these fields is now understood not only as a "science" to learn about nature but also as a "technology" of susceptible alteration. The intersection between biotechnology and bioengineering and its kindred disciplines proved their economic importance, being capable to expand and promote the manufacture of products and services, besides modifying processes in favor of human benefit [6–8].

The current scenario points for biotechnology and bioengineering as being the main technology of the twenty-first century should be absorbed by the general public. Undoubtedly, knowledge of the principles of vital processes already achieved will proportionate changes significant in the society. Therefore, it is important to

IntechOpen

ensure a broad awareness of what these two fields of knowledge involve and what the consequences of accepting or rejecting the innovations [9].

Thus, the chapters presented in this book are intended to help provide a deeper understanding about the recent progresses on biotechnology and bioengineering contributing substantially to the consolidation of bio-based processes and products.

Author details

Rosangela Rodrigues Dias, Leila Queiroz Zepka and Eduardo Jacob-Lopes*
Food Science and Technology Department, Federal University of Santa Maria, UFSM, Santa Maria, RS, Brazil

*Address all correspondence to: ejacoblopes@gmail.com

References

[1] IUPAC. Compendium of Chemical Terminology. Oxford, UK: Blackwell Scientific Publications; 2014. DOI: 10.1351/goldbook.B00666

[2] Munshi A, Sharma V. Safety and ethics in biotechnology and bioengineering: What to follow and what not to. In: Omics Technologies and Bio-Engineering. London, UK: Academic Press; 2018. pp. 577-590. DOI: 10.1016/B978-0-12-804659-3.00025-7

[3] Amarakoon II, Hamilton C-L, Mitchell SA, Tennant PF, Roye ME. Biotechnology. In: Pharmacognosy: Fundamentals, Applications and Strategies. London, UK: Academic Press; 2017. pp. 549-563. DOI: 10.1016/B978-0-12-802104-0.00028-7

[4] Mustafa MG, Khan MGM, Nguyen D, Iqbal S. Techniques in biotechnology: Essential for industry. In: Omics Technologies and Bio-Engineering. London, UK: Academic Press; 2018. pp. 233-249. DOI: 10.1016/B978-0-12-815870-8.00013-9

[5] Crawford CA. Principles of Biotechnology. Ipswich, Massachusetts: Salem Press; 2018. pp. 1-466

[6] Gartland KM, Gartland JS. Opportunities in biotechnology. Journal of Biotechnology. 2018;**282**:38-45. DOI: 10.1016/j.jbiotec.2018.06.303

[7] Tylecote A. Biotechnology as a new techno-economic paradigm that will help drive the world economy and mitigate climate change. Research Policy. 2019;**48**(4):858-868. DOI: 10.1016/j.respol.2018.10.001

[8] Singh SP, Pandey A, Du G, Kumar S. Current Developments in Biotechnology and Bioengineering: Synthetic Biology, Cell Engineering and Bioprocessing Technologies. Amsterdam, Netherlands: Elsevier; 2019. pp. 1-442. DOI: 10.1016/C2017-0-01925-0

[9] Farouki N. Biotechnologies: The ideal victim? Advances in Botanical Research. 2018;**86**:325-336. DOI: 10.1016/bs.abr.2017.11.008

Chapter 2

Steps and Tools for PCR-Based Technique Design

Nelson Enrique Arenas and Luz Mary Salazar

Abstract

The identity and clonal differences within bacterial populations have been broadly explored through PCR-based techniques. Thus, bacterial identification and elucidation of DNA fingerprinting have provided insights regarding their phenotypic and genotypic variations. Indeed, some diversity of rates may reflect changes among subpopulations that have their own ecological dynamic and individual traits on coexisting genotypes. Therefore, identification of polymorphic regions from nucleic acid sequences is based on the identification of both conserved and variable regions. Advantages of PCR-based methods are high sensitivity, specificity, speed, cost-effectiveness, and the opportunity for simultaneous detection of many microbial agents or variants. Fingerprint information might allow the tracking of certain outbreaks globally in several reference databases containing valuable genotyping information. In this chapter, we will review applications from Web resources and computational tools online for the designing of PCR-based methods to identify bacterial species. We will also focus on lab applications and key conditions for technique standardization.

Keywords: PCR, molecular biology, diagnosis, molecular typing, DNA fingerprinting, bacterial identification, DNA amplification

1. Introduction

Bacterial culture is the conventional test to identify a microorganism which is based on the isolation and growth of live specimens [1]. Currently, these methods have been considered the gold standard method for assessing the validity of new diagnostic methods. However, phenotypic-based methods are time-consuming tests, difficult to interpret, low reproducible between laboratories, expensive, and laborious. Several commercial polymerase chain reactions or PCR-based methods are available nowadays, and like *in-house* PCR assays, they use different target specific genes in a clinical sample to identify a pathogen [2].

Bacterial identification in many cases is performed through a fingerprint comparison against some reference genotyping databases. Thus, many organisms can be taxonomically classified and specifically differentiated according to several conserved genes which are recognized as "molecular clocks" [3]. One well-known example is the ribosomal RNA (rRNA) gene, which is a good candidate due to its universal distribution and reasonably well conservation in sequence across evolution [4]. Thus, a good method is feasible, rapid, and cheap and can be implemented in local settings from highly endemic areas of a certain infectious disease [5]. Based on designing a proper PCR-based method, the challenge was addressed on serial steps to reach the expected aim.

2. Step one: choosing the PCR-based method

The first step is to select an accurate technique that provides enough genetic information about your model organism. Therefore, it is pivotal to outline a molecular method with high reproducibility, specificity, sensibility, and discrimination power. Some nucleic acid-based techniques with similar taxonomic ranges as other fingerprinting techniques for strain characterization include: restriction fragment length polymorphism (RFLP), pulsed-field gel electrophoresis (PFGE), amplified fragment length polymorphism (AFLP), random amplified polymorphic DNA (RAPD), multilocus sequence typing (MLST), arbitrarily primed PCR (AP-PCR), repetitive sequence-based PCR (rep-PCR), and internal transcribed spacer-PCR (ITS-PCR) (**Table 1**). These techniques are regularly used for classification at genus, species, and subspecies level and even for strain characterization [6].

Molecular biologists are privileged to have a repertoire of tools which provide good molecular distinction which can resolve questions in the clinical settings and infectious disease control. Those methods are regularly performed for pathogen characterization and gave higher positive detection of target species than conventional methods [7]. Therefore, interlaboratory reproducibility is an important feature to produce highly valuable information [8, 9]. Thus, genotyping would provide accurate data that allow the implementation of multi-user international libraries, for example, MLST for multiple bacterial species and spoligotyping for mycobacteria [10–12]. Surveillance studies have identified certain species through DNA fingerprinting which can acquire drug-resistance determinants or clones prone to global dissemination [13, 14]. Tracking certain genotypes is a key to control them in specific geographic areas [15].

Techniques based on DNA amplification include two main approaches: the first involves amplification of one or few fragments of specific regions followed by specific restriction (e.g., ARDRA). The new fragments can be resolved by gel electrophoresis which provides specific patterns for each strain. Besides, differences in the DNA sequences or genomic regions allow its identification or polymorphisms specific at genus, species, or strains level [6]. On the other hand, unspecific probes or oligonucleotide arrays can be used for multiple amplification into the chromosomal DNA. Fragments are also separated by gel electrophoresis resulting in a band pattern or fingerprint that represents a specific genotype. Some available protocols included: arbitrary primers (e.g., RAPDs), double digest selective label (DDSL), extended primers in combination with low annealing temperatures (e.g., AP-PCR), or those based on DNA repetitive sequences from several bacteria genus (e.g., rep-PCR).

Family	Genus	Species	Subspecies	Strain
		Whole-genome sequencing		
	Sequencing of 16S rRNA genes			
	ARDRA, tRNA-PCR, DNA-DNA reassociation			
		RFLP, PFGE, AFLP, RAPD, MLST, AP-PCR, rep-PCR, ITS-PCR		

ARDRA, amplified rDNA (ribosomal DNA) restriction analysis; tRNA-PCR, PCR-amplified length polymorphisms in tRNA intergenic spacers; RFLP, restriction fragment length polymorphism; PFGE, pulse-field gel electrophoresis; AFLP, amplified fragment length polymorphism; RAPD, random amplification of polymorphic DNA; MLST, multilocus sequence typing; AP-PCR, arbitrarily primed PCR; rep-PCR, repetitive sequence-based PCR; and ITS-PCR, internal transcribed spacer-PCR.

Table 1.
Techniques commonly used for bacterial genotyping and identification [6].

3. Step two: defining a gene target

To analyze the genetic variability, it is necessary to select an appropriate target which supports the required hierarchical level. Important insights remain regarding the choice of a "molecular clock" including traces of the evolutionary record from microbial diversity [16]. Ribosomal genes become ancient molecules that harbor information with high phylogenetic value and can differentiate organisms at genus and species level [17]. However, alternative core genes also defined as housekeeping have been proposed such as RNA polymerase beta subunit (*rpoB*), DNA gyrase alpha subunit (*gyrA*), glyceraldehyde 3-phosphate dehydrogenase A (*gapA*), GroEL genes (*groE, groL*), outer membrane protein A (*ompA*), and glucose-6-phosphate isomerase (*pgi*) ensuring the success for bacterial species definition [18–20]. Some of these genes are included in a multi-locus sequence typing (see further information in step ten).

Typing methods based on the 16S rRNA genes represent an accurate strategy for strain characterization because these genes harbor both conserved and variable regions that might delineate changes on a specific position on the bacterial ribosome leading to strain differentiation [4, 17]. Although, it is important to consider that multiple copies of the 16S rRNA genes are present in all bacterial genomes. 16S rRNA gene has been subjected to many phylogenetic studies including those related to bacterial definition species [17].

Sequencing of rRNA genes is the preferential method for phylogenetic reconstruction, nucleic acid-based detection, and quantification of microbial diversity [21]. Thus, many genotyping approaches remain based on 16S rRNA gene analysis or ribosomal gene sequencing which still constitutes a gold standard for bacterial taxonomy [22]. Therefore, it is possible to explore a sequence by searching against the Ribosomal Gene Database (RGD) release 11.5 which contains 3.356.809 16S rRNAs and 125.525 fungal 28S rRNA gene sequences by November on 2018 (https://rdp.cme.msu.edu/) [23, 24]. A similar resource containing information regarding the ribosomal genes is the SILVA databases (https://www.arb-silva.de/). It includes over 6,800,000 small (16S/18S, SSU) and large (23S/28S, LSU) subunit rRNA sequences [25]. Another useful database provides detailed information about the ribosomal protein gene database (RPG). This tool contains information from some eukarya, archaea, and bacteria organisms including sequences (genomic, cDNA, and amino acid sequences), intron/exon structures, genome locations, small nucleolar RNAs (snoRNAs), and ortholog data (http://ribosome.med.miyazaki-u.ac.jp) [26].

4. Step three: primer design for PCR-based methods

Once we have chosen a gene target, the next step is the design of specific primers to detect it in the DNA sample. In the last decades, PCR-based techniques have been successfully employed for the genetic characterization of many taxa of many pathogens [2, 27]. Primers for DNA amplification are short synthetic oligonucleotides which may be complementary to target sites on the template DNA. PCR is performed at different temperatures (denaturation, annealing, and extension) where efficiency is determined based on primer annealing [28]. Some essential features have to be taken into consideration for accurate primer design:

1. Primers should contain guanine or cytosine, or both at the 3′-ends to increase the efficiency of oligonucleotide binding. Primers must form a stable duplex with target DNA at the annealing temperature.

2. Oligonucleotide should not be self-complementary to avoid the generation of secondary structures such as hairpins loops.

3. No complementarity between forward and reverse primers.

4. Melting temperature (T_m) defines the balance between the unbounded primer and free template compared with primer bound to the target DNA (50%). Tm is recommended to be among 42–65°C with an ideal temperature of 62°C. This parameter is a key because too high T_m can affect specificity and decrease the amount of PCR product (amplicon). If T_m is too low, unspecific products can be produced due to mismatch base pairing. Usually, T_m is set, 5°C below or 5°C above.

5. The optimal length of primers should be long enough to increase specificity (usually between 18 and 30 bases).

6. It is recommended to ensure a similar distribution of G/C and A/T content in the primers (40–60%).

7. For standard PCR, length of the amplified product should be between 200 and 1000 bp and for quantitative PCR in the range from 75 to 150 bp. Nevertheless, PCR products larger than 1000 bp may require additional time during the extension step (1 min/kb of PCR product).

8. Primers must be designed specifically in the target gene to avoid cross contamination with unwanted DNA sequences in the PCR. Typically, primers are designed, and sequences are analyzed *in silico* using BLAST analysis or others, to check for the specificity.

Many resources are available for primer and probe design to optimize the PCR method; therefore, the researcher might consider parameters including molecular weight, millimolar extinction coefficient, T_m, and prediction of secondary structure formation and magnesium chloride ($MgCl_2$) concentration (**Table 2**). For some molecular biology procedures, it is recommended to design the forward primer less 35 pb downstream from the start site of the coding gene and also it applies for the reverse primer regarding the stop site. For example, in a sequencing protocol, the fragment size should not be large due to artifacts introduced on lecture sequence. Likewise, unclear results might increase concomitantly with larger fragments [29].

5. Step four: *in silico* simulation of molecular biology experiments

Nowadays, many resources (Web servers and programs) are available to simulate PCR results, predicting expected bands and successful primer annealing [30, 31]. Although, *in silico* simulation of several PCR-based methods is possible by using tools to obtain theoretical PCR results with many bacterial species sequenced up to date [32]. The list of target genomes is updated according to their availability at NCBI. Many experiments against prokaryotic genomes can be performed such as PCR amplification, restriction digest, and PFGE, PCR-RFLP, double digestion fingerprinting, AFLP-PCR, and other DNA fingerprinting techniques (http://insilico.ehu.es/) [33]. The PCR simulation is also possible if the researcher already knows the target sequence and can test it by using certain resources already mentioned in **Table 2**.

Resource	Description
Oligo®.net	This primer analysis software is useful to design sequencing and PCR primers. Also, user can check DNA and RNA secondary structure, dimer formation, false priming, and homology. Available at https://www.oligo.net/
CLC Main Workbench	It is commercial software from Qiagen. It provides basic and advanced sequence analysis in a user-friendly platform. Access at: https://www.qiagenbioinformatics.com/products/clc-main-workbench/
Primer Premier	It is a PCR Primer Design Software from Premier Biosoft company. Primer Premier's search algorithm finds optimal PCR, multiplex and SNP genotyping primers with the most accurate melting temperature using the nearest neighbor algorithm. Available at: www.premierbiosoft.com/primerdesign/
Gene Runner	It is a free program for multiple sequence analysis including PCR, restriction analysis, sequencing primers, and hybridization probes. Available at: http://www.generunner.net/
Serial Cloner	Serial Cloner 2.5 (free) lets the user to find restriction sites, define ORF, calculate T_m of selected fragments, %GC, and scan sequence features. Download at: http://serialbasics.free.fr/Serial_Cloner.html
Primer designing tool	A tool for finding specific primers on a PCR template developed by the National Center for Biotechnology Information (NCBI), U.S. Available at: https://www.ncbi.nlm.nih.gov/tools/primer-blast/
Primer3web version 4.1.0	Since the first release, this popular Web interface has assisted many researchers with PCR primer design. The newer version of Primer3 available at http://primer3.ut.ee
Genefisher2	Genefisher2 is an interactive Web-based program for designing degenerate primers. Access at: https://bibiserv.cebitec.uni-bielefeld.de/genefisher2/
PrimerExplorer	This software is specifically for designing the primer sets for loop-mediated isothermal amplification (LAMP) method. https://primerexplorer.jp/e/
Tools for calculation of oligo properties	The following tools calculate T_m, molecular weight, and millimolar extinction coefficient (OD/μmol, μg/OD) for oligos: Oligo Calc: Oligonucleotide Properties Calculator available at http://biotools.nubic.northwestern.edu/OligoCalc.html Oligo Calculator http://mcb.berkeley.edu/labs/krantz/tools/oligocalc.html Oligo Calculation Tool https://www.genscript.com/tools/oligo-primer-calculation
Interactive tools provided by suppliers	New England Biolabs with multiple tools available at https://www.neb.com/tools-and-resources/interactive-tools. Thermo Fisher scientific provides the Oligo Design Tools which can be found at: https://www.thermofisher.com/co/en/home/life-science/oligonucleotides-primers-probes-genes/custom-dna-oligos/oligo-design-tools.html Eurofins offers a PCR Primer Design Tool at: https://www.eurofinsgenomics.eu/en/ecom/tools/pcr-primer-design/ Sigma-Aldrich provides the OligoEvaluator™ http://www.oligoevaluator.com/ShowToolServlet?TYPE=ANALYSIS OligoAnalyzer 3.0 from Integrated DNA Technology (IDT): https://www.idtdna.com/calc/analyzer GenScript Real-time PCR (TaqMan) Primer Design https://www.genscript.com/tools/real-time-pcr-tagman-primer-design-tool
Directory of different tools	This directory contains links for: calculation of oligonucleotide physicochemical parameters, PCR primers based upon protein sequence, PCR and cloning, PCR primers based upon multi-alignments, genomic scale primers, overlapping primer sets, short interfering RNA (siRNA) design, real-time PCR primer design, introduction of mutations, and primer presentation on the DNA sequence. Available at: https://molbiol-tools.ca/PCR.htm

Table 2.
Available tools for primer design.

6. Step five: guidelines for standardization of PCR mix

Once, you have your DNA target and PCR primer sequences, you should set experimental conditions including water (HPLC-grade), 10× reaction buffer, $MgCl_2$, dNTPs, primers (forward and reverse), sample DNA, and DNA polymerase [29]. The 10× reaction buffer includes magnesium, thus it is optional to use separately as $MgCl_2$. If so, typical $MgCl_2$ concentration in a standard PCR should be between 1.5 and 2.0 millimolar (mM). When magnesium is too low, no amplicon might appear; but if it is too high, undesired amplicons would be observed as extra bands in the agarose gel. For dNTP, concentration should be 200 μM of each nucleotide. Regarding the Taq (obtained from *Thermus aquaticus*) DNA polymerase (further information in step 6), it is recommended the addition between 0.5 and 2.0 units per 50 μL mix (preferably 1.25 units). Primers work well at the default concentration (50 nM), but concentrations between 0.1 μM and 1 μM of each primer are recommended. Last, DNA sample should be used between 1 ng and 1 μg of genomic templates because higher amounts might reduce PCR product specificity [28].

7. Step six: choosing the DNA polymerase enzyme

DNA polymerase is an enzyme which synthesizes the new DNA strands. DNA polymerase was first isolated from *T. aquaticus* in 1976 [34]. This enzyme has an optimum temperature between 75 and 80°C, which possess a half-life until 97.5°C during 9 min and can polymerize 150 nucleotides per second [35]. When choosing a DNA polymerase, the researcher must consider key aspects such as specificity, thermostability, fidelity, and processivity. First, if specificity is low, low-quality amplicon would affect the yield product, sensitivity, and possible problems in downstream applications (e.g., cloning or protein expression). Second, regarding thermostability, consider using enzymes with a half-life above 90°C because of the denaturing step. Third, if the researcher needs amplicons with 100% similarity to DNA target, consider using high-fidelity DNA polymerases with significantly proof-reading activity. Last, processivity reflects the rate and speed of the reaction from the enzyme. Thus, processivity should be considered in case of long templates, self-complementary targets, high G/C content, and samples containing PCR inhibitors including if the amplicon is accumulating during later PCR cycles [29, 36].

8. Step seven: setting the PCR conditions

The PCR runs in cycles composed of three called steps: denaturation, annealing, and extension. For default, PCR includes between 25 and 35 cycles per reaction. The denaturation step produces single-stranded DNA and usually is performed initially at 95°C for 2 min [37]. The following step is the primer annealing which pair-base primer with the complementary DNA template and generally is carried out considering the primer's T_m of both PCR oligos. Third, for extension step is recommended to set 1 min per 1000 bp (or 30 s per 500 bp) of the amplicon. Larger PCR products (>3 kb) may require longer extension times. The extension is usually performed at 72°C which is considered the optimum temperature for thermostable DNA polymerases. The standard PCR protocol for a 500 bp amplicon includes: an initial denaturation at 95°C for 2 min, followed by 25 cycles of denaturation at 95°C for 15 s, annealing at 55°C for 15 s, and extension at 68°C for 45 s. One additional cycle of final extension is at 68°C for 5 min [28, 29].

9. Step eight: setting specificity and sensitivity of the PCR method

To test assay specificity, it should be assessed against many related microorganisms. Potential cross-reactivity with DNA contaminants in the sample should also be investigated especially when the method applies to natural populations [38]. This issue is essential, particularly when the new method is compared with traditional techniques.

Specificity is first tested *in silico* using the BLAST tool [39]. Then, specificity can be assessed in vitro by the PCR amplification of genomic DNA purified from taxonomically related species. Regarding sensitivity, defines the detection limit of the minimum of DNA target in a sample. This issue is relevant when it is difficult to obtain cultures or when the low number of bacteria cannot be detected in other diagnostic technique [40].

Identifying the bacterial species related to clinical phenotypes requires a method to cluster fingerprints into groups which are likely to share most genotypic and phenotypic traits [8]. Instead, species have been genotyped by measuring genetic variation in the number of repetitive genetic elements in the direct repeat region or detection of polymorphic sequence [41]. These techniques have identified clonal groups of isolates that each appears to be related through a common ancestor [42].

10. Step nine: evaluation of the amplicons

Experimental validation of PCR results entails two possibilities: the first option is to load and run the PCR product on an agarose gel testing the expected sizes with the suitable molecular weight marker. In the second place, it is to sequence the amplicon evaluating its sequence identity compared to the DNA target [28, 29].

11. Step ten: comparing a query band pattern or DNA sequence against databases

Genotyping programs rely on the collection and analysis of large quantities of data. Control infection programs are implementing genotyping programs for comparing against a database. Central databases for isolate tracking, laboratory

Resource	Description	Reference
PubMLST	This database contains more than 140 MLST allelic profiles and sequences. BIGSdb software runs in PubMLST to store and analyze sequence data for bacterial isolates. https://pubmlst.org/	[10]
MLST 2.0	Multilocus sequence typing from an assembled genome or from a set of reads. It is brought by the Center for Genomic Epidemiology https://cge.cbs.dtu.dk/services/MLST/	[45]
LOCUST	It is an automated classification program that allows users to customize the typing of microbial isolates from whole-genome sequencing data. http://sourceforge.net/projects/locustyper	[47]
BacWGSTdb	This database was designed for genotyping and source tracking bacterial pathogens. http://bacdb.org/BacWGSTdb/	[15]
PrimerBank	Databank of PCR primers for gene expression detection and quantification (real-time PCR). User can search by using GenBank Accession, NCBI protein accession, NCBI Gene ID, Gene Symbol, PrimerBank ID or Keyword (gene description) or blast a gene sequence. https://pga.mgh.harvard.edu/primerbank/	[50]
RTPrimerDB	It is a public database for primer and probe sequences used in real-time PCR assays. http://www.rtprimerdb.org/	[48, 49]
RUCS 1.0	Rapid identification of PCR primers pairs for unique core sequences. https://cge.cbs.dtu.dk/services/rucs/	[46]

Table 3.
Web resources for genotyping based on MLST and WGS data. Primer databases were also included.

results, and epidemiologic data are essential. Because cluster investigations are an epidemiologic activity, the infectious disease programs should maintain the principal databases for spread analysis and control measurements [43]. The information in these databases can enable infectious disease programs to identify easily patients with matching genotypes and epidemiologic links [44]. Today, information available regarding bacterial genotyping at both traditional MLST and whole-genome sequencing (WGS) are available (**Table 3**) [10, 15, 45–47]. For more applications, public resources also store primer information for quantitative gene expression analysis or comparing with previous reports [48–50].

12. Variants of the PCR methods

Beyond PCR use in the lab, availability of an improved version of thermocyclers, dyes, primers, probes, and DNA polymerases have extended applications.

Technique	Principle	Use	Reference
In house PCR	Conventional PCR targeting a single gene	Detection of a pathogen in a clinical sample	[62]
Nested PCR	Two pairs of primers are used to amplify a fragment. First pair amplifies similar to a conventional PCR. Second pair is nested within the first fragment	Increase specificity of an amplification reaction when targeting a gene	[61]
Multiplex PCR	This PCR variation enables the simultaneous amplification of many targets in a single reaction by using over one pair of primers. Amplicon sizes should be different or be labeled	Screening for a set of genes at once in a DNA sample. Analysis of microsatellites and SNPs	[60]
Real-time PCR or quantitative PCR	It is an assay that monitors the fluorescence emitted during the reaction as an indicator of amplicon production at each PCR cycle (real-time) as opposed to the endpoint detection	DNA quantification in a sample. Level of gene expression. Copy number variation. Genotyping. Multi-species analysis	[59]
Inverse PCR	In this PCR, primers are oriented in the reverse direction of the usual orientation. The template for the reverse primers is a restriction fragment that has been self-ligated	Cloning of sequences flanking a known sequence. Amplification and identification of flanking sequences	[58]
Arbitrary Primed PCR (AP-PCR)	This technique uses primers whose nucleotide sequence is randomly chosen. Amplification occurs under low stringency conditions	DNA fingerprinting	[55–57]
Reverse transcriptase PCR	This PCR transcribes RNA into DNA. It allows the synthesis of complementary DNA (cDNA)	Evidence regarding the transcription of a gene. Complementary DNA synthesis	[51]
Loop Mediated Isothermal Amplification (LAMP)	This PCR variant runs at a single temperature (usually 63–67°C) and requires 6 primers, and it is amenable to visual detection. Amplification occurs in less than 30 min	Point of care methods for pathogen screening	[52]
Droplet PCR	It includes a separation step of sample into multiple compartments so that only few molecules are present in each partition. Thus, each droplet will be an independent PCR	Rare species detection and mutations with low frequency	[53]

Table 4.
Comparison of several PCR-based methods.

Many PCR variants allow quantify gene expression, improving diagnostic sensitivity and genotyping without further procedures as restriction analysis (**Table 4**) [51–64]. Advanced technique such as digital PCR (dPCR) has showed to improve sensitivity and reliability until single-cell applications and tolerance to PCR inhibitors such as chelating agents [64]. Finally, another common application of several PCR-based techniques (conventional PCR, RT-PCR, qPCR, and LAMP assays) is the detection of antimicrobial resistance against first-line drugs or even last-resort antibiotics [65].

13. Educational tools for beginners

Biotechnology students face challenges understanding molecular biology principles and techniques. E-learning resources simulate a real environment where students create a real experience when running protocols in the lab [66, 67]. The genetic science learning center provides a flash-simulation of PCR principles useful for beginners (https://learn.genetics.utah.edu/content/labs/pcr/). This virtual lab explains step-by-step concepts behind the technique principles and concept background. Another educational tool to train students not only in the method but also regarding interpretation results is the virtual application for bacterial identification (available on http://www.hhmi.org/biointeractive/vlabs/bacterial_id/index.html). The experiment's purpose is to use molecular tools to identify different bacterial species based mainly on their 16S rRNA genes. This virtual lab also requires that students prepare a patient's sample, isolate the whole bacterial DNA, perform a DNA sequencing analysis, run the query sequence in the BLAST tool, and identify the pathogen. Finally, students can practice PCR-RFLP simulations through online exercises based on available bacterial genomes [32].

14. Case study: identification at species and subspecies level

PCR-based detection based on the conserved regions of the 16S rRNA sequence of bacterial pathogens is currently performed by several groups [4, 18, 21]. The rRNA at SSU contains segments that are conserved in species, genus, and kingdom level. In this case, *Klebsiella pneumoniae* is divided into three subspecies: *K. pneumoniae pneumoniae*, *K. pneumoniae rhinoscleromatis*, and *K. pneumoniae ozaenae*. All together are phenotypically closely related and difficult to differentiate based on conventional tests [68].

The 16S rRNA gene sequences from *K. pneumoniae* subspecies were retrieved in FASTA format and aligned by using MACAW program (download link http://en.bio-soft.net/format/MACAW.html). Subsequently, sequences were virtually cleaved according to restriction enzyme database obtained from Rebase (GCG format) implemented in GeneDoc program (available in https://www.softpedia.com/get/Science-CAD/GeneDoc.shtml). Restriction patterns were predicted by testing each enzyme and considering specific patterns. Primers were custom-designed using Gene Runner program targeting 16S rRNA gene (**Table 2**). Experimental validation was performed by harvesting the reference strains and extracting their genomic DNA. After PCR, amplification products and restriction enzyme digests were resolved by using agarose gel electrophoresis. Restriction enzyme combinations generated fragments that allowed easy identification of subspecies after separation of digested DNA. Primer design entailed a challenge since 16S rRNA genes are highly conserved in all bacterial genomes and may be present in multiple copies [69, 70]. PCR-RFLP protocol was designed specifically to detect single nucleotide polymorphisms in the 16S rRNA genes from *K. pneumoniae* subspecies. In conclusion, our method combined the detection of specific nucleotide polymorphisms with the easy identification and categorization of the three bacterial subspecies from *K. pneumoniae* [68].

In another case, it was the implementation of *in-house* PCR method for tuberculosis diagnosis [62]. This molecular method might be an important tool in high-incidence areas due to its speed, sensitivity, and discriminatory power overcoming conventional methods (acid-fast stain and culture). The method was designed based on the *IS6110* gene specific for *Mycobacterium tuberculosis* complex and was successfully tested in sputum, bronchoalveolar lavage fluid, blood, gastric fluid aspirate, urine, cerebrospinal fluid, ascitic fluid, and abscess secretions. The method improved diagnostic accuracy and confirmed to be fast, low cost, and feasible and can be implemented in a middle-income resource setting [62].

15. Conclusions

Some bacterial pathogens may be undetectable by traditional culture methods due to their nutrient requirements, growth conditions or the bacterial inoculum per sample. Therefore, PCR emerged as the effective method which overcomes the detection limit of certain pathogens in clinical samples. The success in the PCR experiment implies planning and results prediction with available tools and resources. Web-based tools and programs are useful for primer design, calculating accurate thermodynamic and physicochemical parameters, changing the thermal cycling protocol, and performing a good experimental design. Success in the PCR-based protocol depends on performing an accurate *in silico* simulation which would allow the optimal selection of reagents and test conditions and to avoid troubleshooting on inefficient reactions. Recommendations in this chapter might enable the researcher to customize and troubleshoot a wide variety of PCR-based methods. Hence, PCR remains as a versatile technique in molecular biology that allows changes in adjustable standard protocol to any gene target choosing the most suitable option for pathogen identification.

Acknowledgements

We would like to acknowledge the Office for Research Funding from Universidad de Cundinamarca.

Conflict of interest

The authors declare not to have any conflict of interests.

Author details

Nelson Enrique Arenas[1*] and Luz Mary Salazar[2]

1 Faculty of Agricultural Sciences, Universidad de Cundinamarca, Cundinamarca, Colombia

2 Chemistry Department, Faculty of Sciences, Universidad Nacional de Colombia, Ciudad Universitaria, Bogotá, Colombia

*Address all correspondence to: nearenass@unal.edu.co

References

[1] Donelli G, Vuotto C, Mastromarino P. Phenotyping and genotyping are both essential to identify and classify a probiotic microorganism. Microbial Ecology in Health and Disease. 2013;**24**(1):20105

[2] Yamamoto Y. PCR in diagnosis of infection: Detection of bacteria in cerebrospinal fluids. Clinical and Diagnostic Laboratory Immunology. 2002;**9**(3):508-514

[3] Ludwig W. Nucleic acid techniques in bacterial systematics and identification. International Journal of Food Microbiology. 2007;**120**(3):225-236

[4] Weisburg WG, Barns SM, Pelletier DA, Lane DJ. 16S ribosomal DNA amplification for phylogenetic study. Journal of Bacteriology. 1991;**173**(2):697-703

[5] Reller LB, Weinstein MP, Petti CA. Detection and identification of microorganisms by gene amplification and sequencing. Clinical Infectious Diseases. 2007;**44**(8):1108-1114

[6] Savelkoul P, Aarts H, De Haas J, Dijkshoorn L, Duim B, Otsen M, et al. Amplified-fragment length polymorphism analysis: The state of an art. Journal of Clinical Microbiology. 1999;**37**(10):3083-3091

[7] Fredlund H, Falk L, Jurstrand M, Unemo M. Molecular genetic methods for diagnosis and characterization of *Chlamydia trachomatis* and *Neisseria gonorrhoeae*: Impact on epidemiological surveillance and interventions. APMIS. 2004;**112**(11-12):771-784

[8] Olive DM, Bean P. Principles and applications of methods for DNA-based typing of microbial organisms. Journal of Clinical Microbiology. 1999;**37**(6):1661-1669

[9] Tyler K, Wang G, Tyler S, Johnson W. Factors affecting reliability and reproducibility of amplification-based DNA fingerprinting of representative bacterial pathogens. Journal of Clinical Microbiology. 1997;**35**(2):339

[10] Jolley KA, Bray JE, Maiden MCJ. Open-access bacterial population genomics: BIGSdb software, the PubMLST.org website and their applications. Wellcome Open Research. 2018;**3**:124

[11] Brudey K, Driscoll JR, Rigouts L, Prodinger WM, Gori A, Al-Hajoj SA, et al. *Mycobacterium tuberculosis* complex genetic diversity: Mining the fourth international spoligotyping database (SpolDB4) for classification, population genetics and epidemiology. BMC Microbiology. 2006;**6**(1):23

[12] Sola C, Filliol I, Gutierrez MC, Mokrousov I, Vincent V, Rastogi N. Spoligotype database of *Mycobacterium tuberculosis*: Biogeographic distribution of shared types and epidemiologic and phylogenetic perspectives. Emerging Infectious Diseases. 2001;7(3):390

[13] Demay C, Liens B, Burguière T, Hill V, Couvin D, Millet J, et al. SITVITWEB—A publicly available international multimarker database for studying *Mycobacterium tuberculosis* genetic diversity and molecular epidemiology. Infection, Genetics and Evolution. 2012;**12**(4):755-766

[14] Pearce ME, Alikhan N-F, Dallman TJ, Zhou Z, Grant K, Maiden MCJ. Comparative analysis of core genome MLST and SNP typing within a European *Salmonella* serovar Enteritidis outbreak. International Journal of Food Microbiology. 2018;**274**:1-11

[15] Ruan Z, Feng Y. BacWGSTdb, a database for genotyping

and source tracking bacterial pathogens. Nucleic Acids Research. 2016;**44**(D1):D682-D6D7

[16] Bromham L, Penny D. The modern molecular clock. Nature Reviews. Genetics. 2003;**4**(3):216

[17] Rajendhran J, Gunasekaran P. Microbial phylogeny and diversity: Small subunit ribosomal RNA sequence analysis and beyond. Microbiological Research. 2011;**166**(2):99-110

[18] Case RJ, Boucher Y, Dahllöf I, Holmström C, Doolittle WF, Kjelleberg S. Use of 16S rRNA and *rpoB* genes as molecular markers for microbial ecology studies. Applied and Environmental Microbiology. 2007;**73**(1):278-288

[19] Gironde S, Manceau C. Housekeeping gene sequencing and multilocus variable-number tandem-repeat analysis to identify subpopulations within *Pseudomonas syringae* pv. *maculicola* and *Pseudomonas syringae* pv. tomato that correlate with host specificity. Applied and Environmental Microbiology. 2012;**78**(9):3266-3279

[20] Koeppel A, Perry EB, Sikorski J, Krizanc D, Warner A, Ward DM, et al. Identifying the fundamental units of bacterial diversity: A paradigm shift to incorporate ecology into bacterial systematics. Proceedings of the National Academy of Sciences. 2008;**105**(7):2504-2509

[21] Janda JM, Abbott SL. 16S rRNA gene sequencing for bacterial identification in the diagnostic laboratory: Pluses, perils, and pitfalls. Journal of Clinical Microbiology. 2007;**45**(9):2761-2764

[22] Clarridge JE. Impact of 16S rRNA gene sequence analysis for identification of bacteria on clinical microbiology and infectious diseases. Clinical Microbiology Reviews. 2004;**17**(4):840-862

[23] Cole JR, Wang Q, Fish JA, Chai B, McGarrell DM, Sun Y, et al. Ribosomal database project: Data and tools for high throughput rRNA analysis. Nucleic Acids Research. 2014;**42**(Database Issue):D633-D642

[24] Wang Q, Garrity GM, Tiedje JM, Cole JR. Naive Bayesian classifier for rapid assignment of rRNA sequences into the new bacterial taxonomy. Applied and Environmental Microbiology. 2007;**73**(16):5261-5267

[25] Quast C, Pruesse E, Yilmaz P, Gerken J, Schweer T, Yarza P, et al. The SILVA ribosomal RNA gene database project: Improved data processing and web-based tools. Nucleic Acids Research. 2012;**41**(D1):D590-D5D6

[26] Nakao A, Yoshihama M, Kenmochi N. RPG: The ribosomal protein gene database. Nucleic Acids Research. 2004;**32**(suppl_1):D168-DD70

[27] Bellizzi D, Sgaramella V. Bacteria identification by PCR-based techniques. Microbial Genome Methods. 1996:167

[28] Innis MA, Gelfand DH, Sninsky JJ, White TJ. PCR Protocols: A Guide to Methods and Applications. Oval road, London, UK: Academic Press; 2012

[29] White BA. PCR Protocols: Current Methods and Applications. Totowa, New Jersey, USA: Springer Science & Business Media; 1993

[30] Rozen S, Skaletsky H. Primer3 on the WWW for general users and for biologist programmers. Bioinformatics Methods and Protocols. Totowa, New Jersey, USA: Springer; 2000. pp. 365-386

[31] Giegerich R, Meyer F, Schleiermacher C, editors. GeneFisher-software support for the detection of postulated genes. In: Proc Int Conf Intell Syst Mol Biol. (ISMB). Vol. 4. 1996. pp. 68-77

[32] San Millán RM, Martínez-Ballesteros I, Rementeria A, Garaizar J,

Bikandi J. Online exercise for the design and simulation of PCR and PCR-RFLP experiments. BMC Research Notes. 2013;**6**(1):513

[33] Bikandi J, Millán RS, Rementeria A, Garaizar J. *In silico* analysis of complete bacterial genomes: PCR, AFLP–PCR and endonuclease restriction. Bioinformatics. 2004;**20**(5):798-799

[34] Chien A, Edgar DB, Trela JM. Deoxyribonucleic acid polymerase from the extreme thermophile *Thermus aquaticus*. Journal of Bacteriology. 1976;**127**(3):1550-1557

[35] Lawyer FC, Stoffel S, Saiki RK, Chang SY, Landre PA, Abramson RD, et al. High-level expression, purification, and enzymatic characterization of full-length *Thermus aquaticus* DNA polymerase and a truncated form deficient in 5′ to 3′ exonuclease activity. PCR Methods and Applications. 1993;**2**(4):275-287

[36] Kainz P. The PCR plateau phase–towards an understanding of its limitations. Biochimica et Biophysica Acta. 2000;**1494**(1-2):23-27

[37] D'Aquila RT, Bechtel LJ, Videler JA, Eron JJ, Gorczyca P, Kaplan JC. Maximizing sensitivity and specificity of PCR by pre-amplification heating. Nucleic Acids Research. 1991;**19**(13):3749

[38] Korbie DJ, Mattick JS. Touchdown PCR for increased specificity and sensitivity in PCR amplification. Nature Protocols. 2008;**3**(9):1452

[39] Altschul SF, Madden TL, Schäffer AA, Zhang J, Zhang Z, Miller W, et al. Gapped BLAST and PSI-BLAST: A new generation of protein database search programs. Nucleic Acids Research. 1997;**25**(17):3389-3402

[40] Yang S, Lin S, Kelen GD, Quinn TC, Dick JD, Gaydos CA, et al. Quantitative multiprobe PCR assay for simultaneous detection and identification to species level of bacterial pathogens. Journal of Clinical Microbiology. 2002;**40**(9):3449-3454

[41] Hulton C, Higgins C, Sharp P. ERIC sequences: A novel family of repetitive elements in the genomes of *Escherichia coli*, *Salmonella typhimurium* and other enterobacteria. Molecular Microbiology. 1991;**5**(4):825-834

[42] Sachdeva P, Virdi JS. Repetitive elements sequence (REP/ERIC)-PCR based genotyping of clinical and environmental strains of *Yersinia enterocolitica* biotype 1A reveal existence of limited number of clonal groups. FEMS Microbiology Letters. 2004;**240**(2):193-201

[43] Urwin R, Maiden MC. Multi-locus sequence typing: A tool for global epidemiology. Trends in Microbiology. 2003;**11**(10):479-487

[44] Sabat A, Budimir A, Nashev D, Sá-Leão R, Van Dijl J, Laurent F, et al. Overview of molecular typing methods for outbreak detection and epidemiological surveillance. Eurosurveillance. 2013;**18**(4):20380

[45] Larsen MV, Cosentino S, Rasmussen S, Friis C, Hasman H, Marvig RL, et al. Multilocus sequence typing of total-genome-sequenced bacteria. Journal of Clinical Microbiolology. 2012;**50**(4):1355-1361

[46] Thomsen MCF, Hasman H, Westh H, Kaya H, Lund O. RUCS: Rapid identification of PCR primers for unique core sequences. Bioinformatics. 2017;**33**(24):3917-3921

[47] Brinkac LM, Beck E, Inman J, Venepally P, Fouts DE, Sutton G. LOCUST: A custom sequence locus typer for classifying microbial isolates. Bioinformatics. 2017;**33**(11):1725-1726

[48] Lefever S, Vandesompele J, Speleman F, Pattyn F. RTPrimerDB: The portal for real-time PCR primers and probes. Nucleic Acids Research. 2008;**37**(suppl_1):D942-D9D5

[49] Pattyn F, Robbrecht P, De Paepe A, Speleman F, Vandesompele J. RTPrimerDB: The real-time PCR primer and probe database, major update 2006. Nucleic Acids Research. 2006;**34**(suppl_1):D684-D6D8

[50] Wang X, Spandidos A, Wang H, Seed B. PrimerBank: A PCR primer database for quantitative gene expression analysis, 2012 update. Nucleic Acids Research. 2012;**40**(D1):D1144-D11D9

[51] Schmittgen TD, Zakrajsek BA, Mills AG, Gorn V, Singer MJ, Reed MW. Quantitative reverse transcription–polymerase chain reaction to study mRNA decay: Comparison of endpoint and real-time methods. Analytical Biochemistry. 2000;**285**(2):194-204

[52] Notomi T, Mori Y, Tomita N, Kanda H. Loop-mediated isothermal amplification (LAMP): Principle, features, and future prospects. Journal of Microbiology. 2015;**53**(1):1-5

[53] Markey AL, Mohr S, Day PJ. High-throughput droplet PCR. Methods. 2010;**50**(4):277-281

[54] Kalendar R, Khassenov B, Ramankulov Y, Samuilova O, Ivanov KI. FastPCR: An *in silico* tool for fast primer and probe design and advanced sequence analysis. Genomics. 2017;**109**(3):312-319

[55] Williams JG, Kubelik AR, Livak KJ, Rafalski JA, Tingey SV. DNA polymorphisms amplified by arbitrary primers are useful as genetic markers. Nucleic Acids Research. 1990;**18**(22):6531-6535

[56] Welsh J, McClelland M. Fingerprinting genomes using PCR with arbitrary primers. Nucleic Acids Research. 1990;**18**(24):7213-7218

[57] McClelland M, Welsh J. DNA fingerprinting by arbitrarily primed PCR. PCR Methods and Applications. 1994;**4**(1):S59-S65

[58] Siebert PD, Chenchik A, Kellogg DE, Lukyanov KA, Lukyanov SA. An improved PCR method for walking in uncloned genomic DNA. Nucleic Acids Research. 1995;**23**(6):1087

[59] Mackay IM. Real-time PCR in the microbiology laboratory. Clinical Microbiology and Infection. 2004;**10**(3):190-212

[60] Alvarez J, Sota M, Vivanco AB, Perales I, Cisterna R, Rementeria A, et al. Development of a multiplex PCR technique for detection and epidemiological typing of *Salmonella* in human clinical samples. Journal of Clinical Microbiology. 2004;**42**(4):1734-1738

[61] Juck D, Ingram J, Prévost M, Coallier J, Greer C. Nested PCR protocol for the rapid detection of *Escherichia coli* in potable water. Canadian Journal of Microbiology. 1996;**42**(8):862-866

[62] Coronado-Ríos S, Arenas-Suárez NE, Gómez-Marín JE. Diagnosis of pulmonary and extrapulmonary tuberculosis using an *in-house* PCR method in clinical samples from a middle-income resource setting. Infection. 2011;**15**:177-183

[63] Ginzinger DG. Gene quantification using real-time quantitative PCR: An emerging technology hits the mainstream. Experimental Hematology. 2002;**30**(6):503-512

[64] Chang C-H, Mau-Hsu D, Chen K-C, Wei C-W, Chiu C-Y, Young T-H. Evaluation of digital real-time PCR assay as a molecular diagnostic tool for single-cell analysis. Scientific Reports. 2018;**8**(1):3432

[65] Anjum MF, Zankari E, Hasman H. Molecular methods for detection of antimicrobial resistance. Microbiology. 2017;**5**:6

[66] Cobb S, Heaney R, Corcoran O, Henderson-Begg S. The learning gains and student perceptions of a second life virtual lab. Bioscience Education. 2009;**13**(1):1-9

[67] Maldarelli GA, Hartmann EM, Cummings PJ, Horner RD, Obom KM, Shingles R, et al. Virtual lab demonstrations improve students' mastery of basic biology laboratory techniques. Journal of Microbiology & Biology Education: JMBE. 2009;**10**(1):51

[68] Arenas NE, Polanco JC, Coronado SM, Durango CJ, Gómez A. Design of a molecular method for subspecies specific identification of *Klebsiella pneumoniae* by using the 16S ribosomal subunit gene. Colombia Médica. 2009;**40**:194-201

[69] Klappenbach JA, Saxman PR, Cole JR, Schmidt TM. Rrndb: The ribosomal RNA operon copy number database. Nucleic Acids Research. 2001;**29**(1):181-184

[70] Klappenbach JA, Dunbar JM, Schmidt TM. rRNA operon copy number reflects ecological strategies of bacteria. Applied and Environmental Microbiology. 2000;**66**(4):1328-1333

Chapter 3

Developments and Perspectives in Bryophyte Biotechnology in Sub-Saharan Africa

Kenneth Yongabi Anchang and Henrik Toft Simonsen

Abstract

The work described here covers an examination of new bioproducts based on sub-Saharan bryophytes. The work includes in vitro testing of extracts from moss and liverworts against plant pathogenic microbes causing food decay and field crop losses. Additionally, we have shown specific antimicrobial activities of *Marchantia debilis* and **moss** against *Erwinia spp* and *Pseudomonas spp*. The extracts were also tested against aflatoxin-producing fungi isolated from food crops such as maize and peanuts. The efficacy of the extracts on clinical dermatological fungal isolates like *Dermatophilus congolensis* has not been reported. This led to the production of an antifungal solution of bryophyte extracts, which was tested in vivo on animals with skin diseases caused by *Dermatophilosis*. Around 99.5% of the animals were treated. The antifungal solution for treatments has been labeled **Bryosol**, while the disinfectants solution is labeled Bryo-disinfectants and the crop-fungicide is labeled Bryo-fungicides. A mini field pilot trial with Bryo-fungicide showed that crops infected with pathogenic fungi were treated. The results provide the first attempt to demonstrate the use of bioproducts for organic treatment of agricultural crops and diseases in animals based on sub-Saharan bryophytes.

Keywords: moss, Marchantia, Bryosol, disinfectants, antifungal, bioproducts, foods, *Dermatophilosis*

1. Introduction

Bryophytes are non-vascular plants, which are constituted of mosses, liverworts, and hornworts [1]. Although not usually seen to have any importance, bryophytes have recently been used as bioindicators of pollution and are often used for decorations. However, the medicinal value of bryophytes is huge with a panoply of bioactive compounds isolated from bryophytes, especially liverworts [1, 2]. Bioactive compounds have been isolated from liverworts from Asia, Europe, and South Africa. For example, Allison in 1975 identified a number of bioactive compounds from liverworts in New Zealand. Volatile constituents have been identified in liverworts like *Tritomaria polita*, *Marsupella emarginata*, *M. aquatic*, and *M. alpine* [3].

It has been seen that bryophytes are rich in diverse phytochemicals such as sesquiterpenoids, norsesquiterpenoids, anthocyanidins, riccionidins etc. with interesting biological activities, such as antimicrobial, antifungal, insect-repellent, molluscicidal, cardiotonic activity, and fragrance compounds among others [1]. Bryophytes are very common across the world, particularly in wet areas like Cameroon. The ecology of Cameroon is rich in algae and lichens; bryophytes in Cameroon are part of the Congo forest and the highlands from Mount Cameroon via the Atlantika Mountains to the Mandara Mountains collectively constituting the Congo forest from Nigeria, ranges from 1400 to 4000 m, and harbors a rich biodiversity of both lower and higher plants. A survey of bryophytes in Cameroon revealed many unidentified species with familiar dormant species such as *Marchantia spp* [2].

Phenanthrenes and other phenolics have been isolated from in vitro cultures of *Marchantia polymorpha.* Recently, extensive report was published on the biology and constituents and chemistry and organic natural products of bryophytes [1, 4], though this lacks data on bryophytes from West and Central Africa, especially Cameroon. Screening of bryophytes and lower plants for biologically active compounds from Cameroon is of great importance considering that Cameroon is centrally placed in Africa and harbors all the ecological and geographical characteristics widespread across Africa. The search for bioactive compounds from plants in the past 30 years in Africa has concentrated on higher plants with little or no interest on bryophytes, which is the same in the rest of the world [2]. The prevalence of many plant and animal pathogenic diseases is growing along with drug resistance strains. This generates huge losses in agricultural yield and productivity across Africa. Treatment and management are expensive for many African farmers and therefore a cheap alternative, preferably organic, is needed [3].

Bioactive compounds from bryophytes could bridge this gap. Here, we show that new drug leads could be identified from bryophytes from Cameroon to address plant pathogenic diseases and animal diseases like *Dermatophilosis* infection in cattle.

2. Dermatophilosis in animals

Dermatophilosis is caused by the bacterium *Dermatophilus congolensis*, which is an aerobic actinomycete (facultatively anaerobic) and usually affects animals and occasionally humans [5].

Dermatophilosis is distributed worldwide, prevailing in tropical areas, and related to humid environments and other factors, such as poor veterinary services, coinfection with a number of bacterial infections, especially in animals with compromised immune systems, and poor hygiene conditions in favor of its occurrence and spread.

In Africa and many other places, the impact caused by animal diseases continues to negatively affect the local economy. Dermatophilosis is a tick-borne disease of ruminants and other animals [5] and affects all parts of the body of the animal. In Nigeria and Cameroon, Dermatophilosis accounts for about 75% of morbidity in herds and about 12% in cattle. Mortality rate has been reported to be quite high due to the resulting toxemia and general debility [5]. Dermatophilosis is an intractable disease and highly contagious, spreading from cattle to man (zoonotic).

The common and orthodox treatment for dermatophilosis is through the use of classical antibiotics like lamstreptocide, charmil, and terramycin long acting (TLA), 1% potassium aluminum sulphate dip, and co-biotic (penicillin and dihydrostreptomycin). Apart from the toxicity of some of these drugs, some of them contain heavy metals, which on accumulation could cause tumors and cancers in both man and animals [3, 6]. The use of organophosphate dips has also been reported to have a negative effect on the environment, and it has been observed to cause systemic damages on internal organs of both animals and humans.

3. Plant pathogenic fungal contamination of food and crops in Africa

Agricultural plant products in sub-Saharan Africa often decay fast due to infection of field crops and harvested products. Most of these plant pathogenic diseases are fungi. Despite the availability of chemicals for control of these pathogens, many farmers find it inaccessible for reasons of costs and lack of adequate knowledge on usage.

In Africa, the predominant food source for more than 70% of the population is grains such as maze and groundnuts. Even though there are new and improved methods for containing these diseases in food crops, there are still great losses due to fungal infections of the crops. A number of reports have shown that aflatoxins producing fungi are predominant with both field and stored maize and groundnuts. Aflatoxin (Aflatoxin B1) is produced by *Aspergillus flavus* or *Aspergillus parasiticus*, and effects of aflatoxin on crops like maize and groundnuts completely destroy the crop due to the toxicity of aflatoxin to humans.

4. Methodology

4.1 Isolation and identification of *Dermatophilus congolensis*

Traditional cultivation techniques were employed for isolation and identification. Swab samples were taken from the lesions on the animal and analyzed at the Phytobiotechnology Research laboratory for *Dermatophilus congolensis* culture. Initial cultures were done in thioglycolate broth and subculture after 48 hours on fortified chloramphenicol Sabouraud Dextrose agar and modified cycloheximide (Actidione)-chloramphenicol Sabouraud agar previously prepared and incubated for 48 hours at 35°C in an aerobic condition. For specific cultural distinct features on agar plate, blood and chocolate agar plates were prepared and distinct colonies from Sabouraud agar plates were transferred on to blood and chocolate agar plates aseptically incubated in air supplemented with 5% CO_2, and the blood agar was also incubated in an anaerobic atmosphere [4, 7].

At 24 hours, a pure culture with tiny, point-like, smooth, creamy, white-colored, beta-hemolytic colonies adherent to the media grew in aerobic blood agar and chocolate agar, with Gram staining showing hypha-like, branching filaments with “train track” forms and clusters of sporangia as well as Gram-positive coccoid forms, mostly in chains. After 48 h, crowded colonies became yellowish and mucoid, with a great variation in colonial morphology, for example, pulvinate, umbonate, or cake crumb-like forms were considered typical of *Dermatophilus congolensis*. This is shown in **Figures 1** and 2.

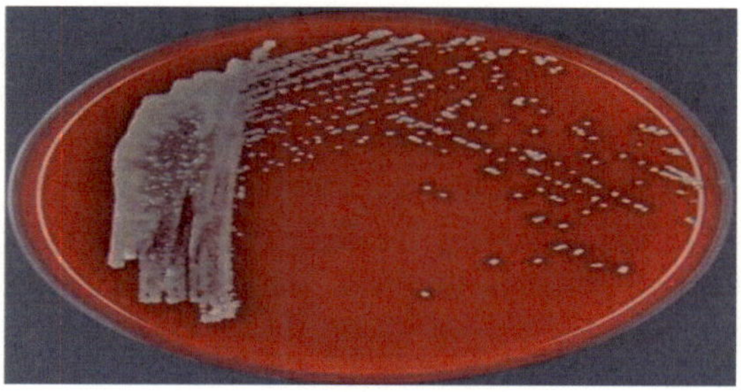

Figure 1.
Beta-hemolytic colonies after 2 days of incubation at 37°C on blood agar medium, with pleomorphic appearance in pulvinate, umbonate, or cake crumb-like form. Dermatophilus congolensis on blood agar plate.

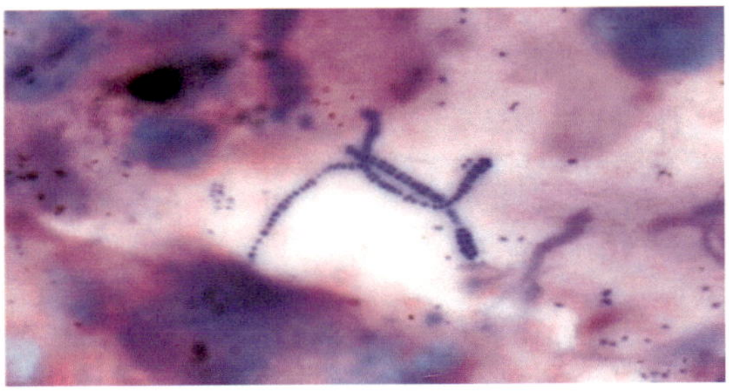

Figure 2.
Gram stain with characteristic branching filaments with "train track" forms or hypha-like chains that released sporangium Gram-positive cells (magnification, 1000×). Beta-hemolytic colonies after 2 days of incubation at 37°C on blood agar medium, with pleomorphic appearance in pulvinate, umbonate, or cake crumb-like form.

Figures 1 and **2** reveal the unique, distinct bacteriological features of *Dermatophilus congolensis.* The biochemical characteristics of *D congolensis* as basis for identification done according Monica Cheesbrough [8] revealed that beta hemolysis in 3–7 days, oxidase, gelatin, casein and starch all positive, while *D congolensis* fermented fructose, ribose and galactose.

5. Survey and extraction of bryophytes

A preliminary survey of liverworts in northwest and southwest regions of Cameroon was performed. Bryophytes (species of liverwort and moss) from Cameroon West/Central Africa were collected and complete sequences for the 18S-rRNA gene of bryophytes were used to construct a phylogenetic tree of bryophytes from Cameroon to fully identify the prevalent species in Cameroon.

5.1 Extraction procedures for the selected and identified bryophyte species

About 50 g of each of the bryophyte (*Marchantia debilis* and Plangiochila spp) plant material were added separately to 250 ml each of methanol and petroleum ether (1:5 w/v) in 250 beakers (Pyrex) for each plant mash and allowed to extract for 72 hours [6]. The extracts were filtered by gravity filtration using Whatman filter paper no 1 locally purchased in Bamenda, Cameroon, and the filtrate solvent was evaporated under vacuum using an incubator at 37°C and the resulting dried extracts were stored in sterile screw-capped bottles and kept at room temperature for further antibacterial testing using extracts of bryophytes. The morphology of the bryophytes and the nature of extracts is shown in **Figures 3–8**.

5.2 Antibacterial activity of the extracts of bryophytes

The agar diffusion method according to Yongabi et al. [9] was employed. Around 0.2 g of the *Marchantia debilis* and *Plangiochila* spp. extracts was reconstituted in 5 ml of distilled water. Antibiotic susceptibility will be determined by agar well diffusion method—commonly used and standardized in the US by National Committee for Clinical Laboratory Standards (NCCLS) [8, 10].

The zone of inhibition was measured and results interpreted as sensitive, intermediate resistant, or resistant. The zone sizes of inhibition were measured and interpreted using the NCCLS as recommended by WHO [8]. Each of the extracts was incorporated in a 6-mm well previously bored using a steel borer. A control set up was established by introducing the extracting solvent (methanol and petroleum) into the different wells as well. The plates were incubated at 37°C for 36 hours. The development of inhibition by the extracts against the test organism was measured [11].

The differences between the inhibition rates of the extracts in the test setup and that of the control were recorded as actual diameter of zones of inhibition caused by the extract. The methanol and petroleum extracts did not exhibit any inhibition in this study, as shown in **Tables 1** and 2.

5.3 Preparation of bryophyte extracts-based ointment using olive oil base

The organic extracts (200 mg each) of *Marchantia* spp. and *Plangiochila* were blended into 200 ml of olive oil and palm kernel oil. Standard organic chemistry protocols as described by Yongabi et al. [9] were applied.

	Fresh *Marchantia debilis*			Dried *Marchantia debilis*		
Microbial test organism	**Hexane extracts**	**Petroleum extracts**	**Methanol extracts**	**Hexane extracts**	**Petroleum extracts**	**Methanol extracts**
Staphylococcus aureus	9 mm	12 mm	14 mm	9.5 mm	17.2 mm	12.5 mm
Pseudomonas aeruginosa	0 mm	0 mm	0 mm	10.5 mm	5.2 mm	12.5 mm
Bacillus spp	0 mm	0 mm	0 mm	11.5 mm	6.2 mm	13.5 mm
Dermatophilus congolensis	5 mm	5 mm	13 mm	6 mm	11 mm	12.5 mm
Aspergillus flavus	3 cm	2 cm	No Growth	2 cm	1 cm	No Growth

Dermatophilus congolensis is an isolate from cow. For Aspergillus flavus (an isolate from maize rot) the inhibition is given as growth of fungi in 7 days, where the control grew 10 cm.

Table 1.
Preliminary in vitro test showing zone of inhibition of organic extracts of Marchantia debilis on different microbes.

	Fresh *Plangiochila spp*			Dried *Plangiochila spp*		
Microbial test organism	**Hexane extracts**	**Petroleum extracts**	**Methanol extracts**	**Hexane extracts**	**Petroleum extracts**	**Methanol extracts**
Staphylococcus aureus	9 mm	8 mm	14 mm	9.8 mm	8.8 mm	15.5 mm
Pseudomonas aeruginosa	0 mm	0 mm	0 mm	0 mm	0 mm	8.9 mm
Bacillus spp	0 mm	0 mm	0 mm	0 mm	0 mm	5 mm
Dermatophilus congolensis	6 mm	6 mm	7 mm	7 mm	8 mm	9.5 mm
Aspergillus flavus	1 cm	1 cm	No Growth	0.5 cm	0.5 cm	No Growth

Dermatophilus congolensis is an isolate from cow. For Aspergillus flavus (an isolate from maize rot) the inhibition is given as growth of fungi in 7 days, where the control grew 10 cm.

Table 2.
Preliminary in vitro test showing average zone of inhibition of organic extracts of Plangiochila spp. on different microbes.

Figure 3.
Marchantia debilis (Liverwort).

Figure 4.
Plangiochila spp (Moss).

Figure 5.
Two samples of fresh Marchantia debilis residues after extraction with hexane and petroleum ether.

Figure 6.
Two samples of partially dried Plangiochila spp residues after extraction with hexane and petroleum ether.

Figure 7.
Partially dried Marchantia debilis extract in hexane and in petroleum ether.

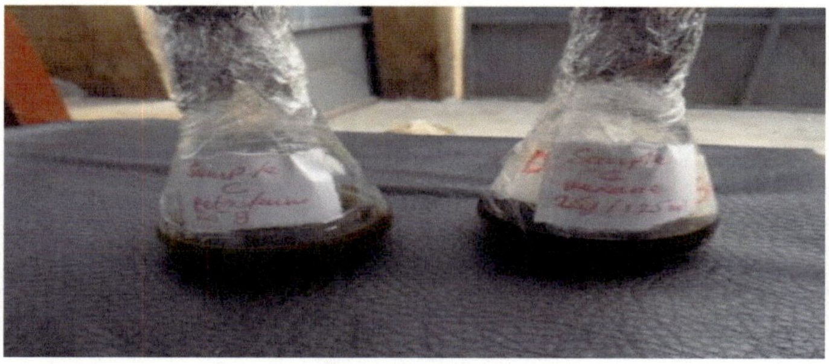

Figure 8.
Extract of Plangiochila spp in hexane and petroleum ether.

5.4 Ointment application and resultant outcome of application

An animal health officer applied the cream topically (by rubbing on affected parts of the animal, using hand gloves) once a day for 3 days in a week. Following this, a total drying off of the infected spot was noticed after 14 days. The dried, dead skin was then carefully peeled off.

6. The findings, discussion on the economic and environmental benefits of this study

The results show that extracts of *Marchantia debilis* and *Plangiochila* have antifungal activity against *Aspergillus flavus*, and antibacterial activities against *Pseudomonas* spp., *Bacillus spp*, and *Staphylococcus aureus* and *Dermatophilus congolensis* isolates (**Tables 1** and **2**). In **Figure 9**, a plate that demonstrates clear mycelia inhibition of *Aspergillus flavus* by extract of *Marchantia debilis* is shown. The product development focus has been on *Marchantia debilis* since Yongabi et al. [2] isolated a number of marchantins including a new marchantin Q from *Marchantia debilis* from Cameroon [2] (**Figures 10** and **11**).

Antimicrobial activity of liverworts is not new [1] but the testing of these liverworts and moss on isolates from plant pathogens in Africa is probably for the first time. The *Aspergillus flavus* isolate was provided for this study by a local laboratory in Cameroon. The effect of liverworts inhibiting the growth of *Dermatophilus congolensis* isolated from cattle is reported here for the first time. The synthetic agrochemicals used in daily farming in Africa are quite expensive and these synthetic products are normally out of reach of the rural farmer [5, 9, 12–14]. The result

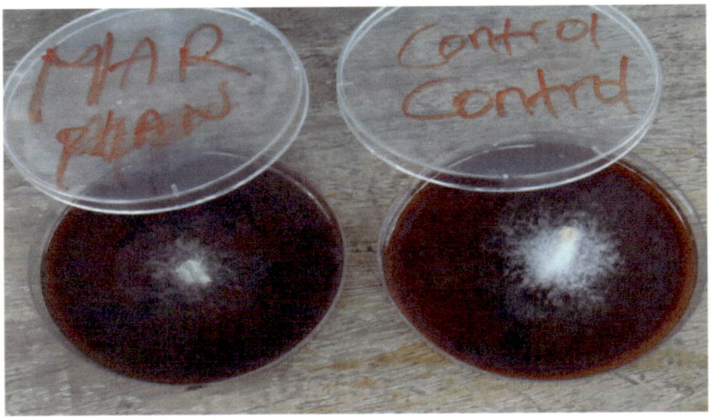

Figure 9.
Plate labeled marked (MAR) contains extract of Marchantia debilis Goebel with slow growth of Aspergillus flavus in 7 days as opposed to control plate (Marked: Control) with only n-hexane incorporated into the agar.

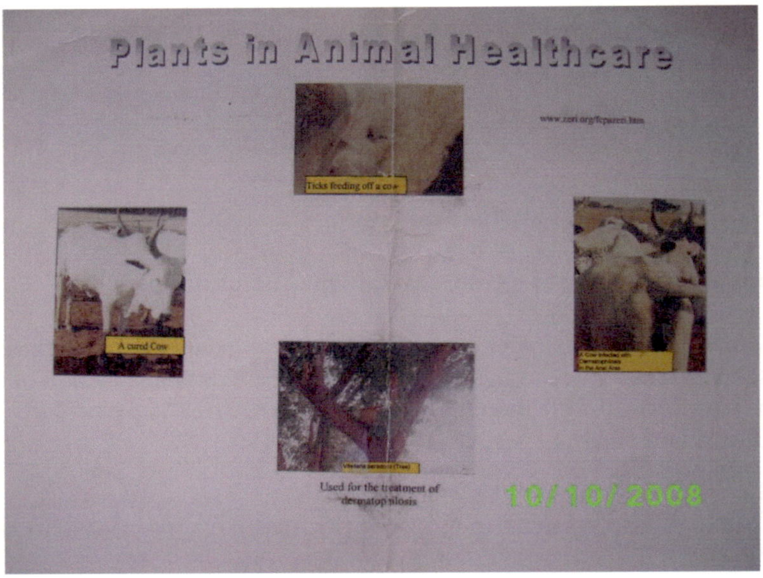

Figure 10.
A cured cow: management of Dermatophilosis in ruminants using Bryo-ointment.

Figure 11.
Maize drying sprayed with Bryo-extracts to inhibit moldy cotton wool Aspergillus flavus.

is that a lot of agricultural produce such as cattle and grains are lost to disease and the cost of these produce is prohibitive. More so because grains are not stored for a longer period to enable sales at off season [15–17].

The problem in cattle with Dermatophilosis is no different. In a lot of the developing countries, today, the problem of malnutrition is endemic and the related opportunistic infections lead to infectious disease, such as tuberculosis and malaria. Protein malnutrition in Africa is a serious problem, especially in rural Africa where approximately 70% of the population live [4, 7, 18, 19]. The chemical constituents of bryophytes are well studied [1, 2], but these rich chemical constituents have not yet been explored biotechnologically. Plants with bioactive ingredients abound in Africa [20], and bryophytes are even more abundant [3].

The production of plant-based products from the bryophytes in the treatment of Dermatophilosis shifts focus from the importation of orthodox drugs and conserves Africa's scarce foreign exchange reserve, and increases utilization of indigenous plant resources. Outside of the cheaper ointment product, a local industry for the production of this ointment is encouraged and the product would be available to a larger group of herders. This preliminary report details the first attempt. The multiplier effect is enormous; meat should be cheaper and malnutrition resulting from the lack of protein would drastically reduce, and possibly disappear. Moreover, the use of plant-based products could easily foreclose the emergent, resistant strains of *Dermatophilus congolensis* resulting from the frequent use and misuse of antibiotics.

Bryophytes are common and abundant, especially in the west and central African regions. The formulation process for the ointment is easy to follow and it is based on a technology that the rural populations could easily handle. The method of application is by glove-protected hands to animals, and the ointment is also effective against human skin infections [9, 14, 21, 22].

The ointment and bryosol (bryophyte solution suspended in glycerol) are observed to have astringent property when applied on sores. Thus, it not only heals but also smoothens the affected lesion to which it is applied. The oils from bryophytes when blended with *Vitellaria paradoxa* have improved cosmetic value and reduced treatment time, and this gives the formulated cream an added advantage. The Bryo-ointment is cheap and effective. For instance, 100 ml is sold at 1 US dollar, as against 5 US dollars for each of the antibiotics. Based on our findings, a container of 100 ml of the formulated cream can hardly be used completely in treating three animals once a day for 3 days a week, and for 2 weeks. However, this depends on the degree of infection.

7. Recommendations

The findings of the study suggest that one cannot begin addressing the problem of aflatoxin producing fungi on crops, grains contamination, and skin diseases in animals by simply relying on agrochemicals and introducing improved management practices. This requires a closer examination of the role of ecological technologies and approaches. Above all, studies on bryophytes are limited to taxonomy and molecular biological aspects with little effort toward actual biotransformation of bryophytes via appropriate biotechnology for direct applications in horticulture and animal husbandry.

From this study, it is therefore recommend that:

- A shift toward cost-effective technology will not take place unless a series of interventions via technology such as bryo biotechnology that can give necessary opportunities is provided to the farmers and other stakeholders.

- Bryophyte bioproducts offer an opportunity for sustainable animal husbandry and agriculture for Africans at potentially lower costs.

- Though farmers pay considerable attention to the selection of seed from their own produce, lack of awareness about identification of contamination in general prevents them from using aflatoxin-free seeds. Interventions such as treating grains with bryophyte-derived solution may ensure that farmers use grains free from contamination irrespective of the sources of supply.

Glossary

Dermatophilosis	bacterial skin disease of cows, sheep, goats, dogs and other animals, scab disease
Amblyomma variegatum	tick vector of Dermatophilosis (cutaneous streptothricosis)
Dermatophilus congolensis	the bacterial causative agent of Dermatophilosis
Kirchi	Hausa name for Dermatophilosis
Lamstreptocide	antibiotics used in treatment of Dermatophilosis
TLA	terramycin long-acting antibiotic used in the treatment of Dermatophilosis
Zoonotic	animal disease that could also be passed to infect man
Necrosis	the breaking down of cells/tissues resulting from an infection
Hausa	the language of Hausa people in Nigeria and Cameroon
Ecology	the study of the relations of living things to one another and to their surrounding

Author details

Kenneth Yongabi Anchang[1,2,3,4*] and Henrik Toft Simonsen[5]

1 Public Health Infectiology and Phytobiotechnology, Imo State University, Owerri, Nigeria

2 Ebonyi State University, Abakaliki, Nigeria

3 Phyto-biotechnology Research Foundation Clinic/Institute, Bamenda, Cameroon

4 Catholic University of Cameroon, Bamenda, Cameroon

5 DTU Bioengineering, Technical University of Denmark, Denmark

*Address all correspondence to: yongabika@yahoo.com

References

[1] Asakawa Y. Chemical Constituents of Bryophytes Progress in the Chemistry of Organic Natural Products. Vol. 95. Singapore: Springer-Verlag wien; 2013. DOI: 10.1007/978-3-7091-1084-3

[2] Yongabi KY, Novakovic M, Bukvicki D, Asakawa Y. Bis-bibenzyls from the Cameroon Liverwort *Marchantia debilis*. Natural Product Communications. 2016;**11**(9):1317-1318

[3] Am A, Paul C, Konig WA, Muhle H. Volatile constituents in the liverwort *Tritomaria polita*. Phytochemistry. 2003;**64**:637

[4] Adam K-P, Becker H. Phenanthrenes and other Phenolics from in vitro cultures of *Marchantia polymorphs*. Phytochemistry. 1994;**35**:139

[5] Abdulkadir IA. Infectious Diseases of Livestock in Nigeria. An Outline. Vol. 274-279. Nigeria: ABU Press Limited; 1989. pp. 37-40

[6] Irobi ON, Daramola SO. Antifungal activities of crude extract of *Mitracarpus villosus*. Journal of Ethnopharmacology. 1993;**40**:137-140

[7] Allison KW, Chud J. The Liverworts of a New Zealand. Dunedin: University of Otugo Press; 1975. p. 300

[8] Cheesbrough M. Medical laboratory manual for tropical countries. Microbiology, Butterworth's. 1984;**2**:76-135

[9] Yongabi KA, Agho MO, Chindo Y, Buba MW. Evaluating the medicinal potentials of some indigenous plants in controlling microbial contamination of poultry feed. Journal of Phytomedicine and Therapeutics. 2000;**5**(2):98-102

[10] Deshpande RG, Khan MB, Bhat DA, Navalkar RG. Inhibition of *Mycobacterium avium* complex isolated from AIDS patients by Garlic. Journal of Antimicrobial Chemotherapy. 1993;**32**:623-626

[11] Moore GS, Atkins RD. The fungicidal and fungistatic effects of an aqueous garlic extracts on medically important yeasts and fungi. Mycologia. 1977;**69**:341-348

[12] Arthur GH, Alen WR. Swine bacterial infections. Equine Veterinary Journal. 1972;**4**(109):15-17

[13] Arthur GH, Noakes D, Pearson H. Veterinary Reproductive and Obstetrics. 5th ed. London: Bailliere Tyndall; 1982. pp. 501-509

[14] Yongabi KA, Agho MO, Chindo IY, Dukku UH. Studies on the antifungal properties of *Urtica dioica*; *Urticaceeae* (stinging nettle). Journal of Phytomedicine and Therapeutics. 2000;**5**(1):39-43

[15] Bida SA, Dennis SM. Dermatophilosis in Northern Nigeria. The Veterinary Bulletin. 1972;**46**:471-478

[16] Lloyd DH, Sellers KC. Dermatophilosis infection in animals and man. In: Edited Papers from 1976, Ibadan Conference. 1976

[17] Railey J, George Mandel H, Sinha S, Judahand DL, Neal GE. Invitro activation of human Harvey-ras Proto Onco gene by aflatoxin b1. Carcinogensis. 1997;**18**:905-910

[18] Dave EL. Chapter 9.26: Aflatoxin toxicology. In: Comprehensive Toxicology. UK: Pergamon Publications; 1997

[19] Groopmann JD, Kensler TW. CRC Critical Reviews in Toxicology. Chapter 19. Ghana: 1999. pp. 113-124

[20] Sofowora A. Medicinal Plants and Traditional in Africa. Part 11. Ife, Nigeria: Pitman Pess Ltd; 1984. p. 128, 142, 146

[21] Njoku CO, Alafiatayo RA. Comparative pathology of the main bovine skin disease in Nigeria. In: A paper presented at the National Conference on Disease of Ruminants, NVRI, Vom. 1984

[22] Ames BN, Profet M, Gold LS. Nature's chemicals and synthetic chemicals. Comparative Toxicology PNAS. 1990;**87**:7782-7786

Chapter 4

Hypotoxic Fluorescent Nanoparticles Delivery by Cell-Penetrating Peptides in Multiple Organisms: From Prokaryotes to Mammalians Cells

Betty Revon Liu, Yue-Wern Huang and Han-Jung Lee

Abstract

Nanotechnology is the study of materials in the nanoscale. By its nature, nanotechnology is interdisciplinary. Nanotechnology has made a significant stride in recent two decades in various industries. Numerous nanomaterials are devised for biomedical applications which include intracellular tracking and labeling, gene detection and hybridization, tumor or tissue targeting, pharmaceutical therapies, pathogenic inhibiting, and medical instrument coating for disinfections. High photostability and quantum yield of fluorescent nanoparticles are ideal for long-term monitoring of molecular events in living organisms. Here, we discuss delivery of three fluorescent nanoparticles in A549 cells, rotifers, Gram-negative bacteria, Gram-positive bacteria, and archaea. As these nanoparticles cannot enter cells, arginine-rich cell-penetrating peptides (CPPs) were used to enhance their internalization at the cellular or organismal level. The 1-(4,5-dimethylthiazol-2-yl)-3,5-diphenylformazan (MTT) assay and sulforhodamine B (SRB) assay demonstrated that CPP complexed fluorescent nanoparticles did not produce lethal effect in all organisms tested. The discussion of these nanomaterials in this chapter intends to broaden our understanding of their biocompatibility in organisms of various hierarchical levels.

Keywords: fluorescent nanoparticles, cell-penetrating peptides, hypotoxicity, rotifer, prokaryotes

1. Introduction

Nanoscience and nanotechnology are fast growing multidisciplinary fields in the past two decades [1]. Nanomaterials are the foundation of devices and systems in various industries that revolutionize functionalities of end-user products [2]. Nanomaterials range from simple zero-dimensional structures such as nanodots [3], wire-like nanocomposites in one-dimension nanoscales [4], to two-dimensional nanosheets, and to three-dimensional structures [5, 6]. Furthermore, anisotropy and unique nano level physical and chemical properties can result in nanomaterials of the same elemental compositions having totally different functionalities [6, 7].

Nanomaterials have contributed to various biomedical applications, including molecular labeling and tracking, DNA/RNA/proteins probing, tumor or tissue targeting, drug delivery and therapies, pathogenic intervention, and biomedical imaging [2, 8–10]. In general, nanomaterials are classified into four categories: carbon-based nanomaterials, metal and metal oxide related nanomaterials, organic-based nanopolymers, and composite nanomaterials with complicated structures [11]. Carbon-based nanomaterials such as fullerenes, carbon nanotubes (CNTs), and graphenes have been used as tissue scaffolds, biosensors, targeted drug delivery, and cosmetic additives [10, 12]. The studies and applications of metal- or metal oxide-nanocomposites are commonly found in toxicology, cancer therapies, and antimicrobial infections [13–15]. Organic-based nanomaterials such as liposomes, micelles, microemulsions, and dendrimers are mainly used in pharmaceuticals and drug delivery systems [16]. These organic nanopolymers can be combined with metal or carbon-based nanoparticles for controlled release of drug delivery and antitumor targeting [11]. There is an alternative classification of nanomaterials based upon their applications: drugs and medications, manufacturing and materials, the environment, electronics, energy harvest, and mechanical industries [17]. The first category can be further divided into cell specificity enhancement [18, 19], drugs, peptides, genes, vaccine delivery [16, 18, 20–25], diagnostics and imaging [2, 24, 26–28], anti-infectious agents or germicides [15, 21, 29], cancer therapy [24, 28–31], tissue/organ/tumor targeting [19, 24–26, 28, 31], scavengers for free radical or thrombosis [32, 33], DNA/RNA/PNA sensor and sequencing [34, 35], and angiogenesis inhibition [36, 37]. Toxicities and health implications become a dominant issue, even though most nanomaterials are hypotoxic for cell/tissue separation or identification, pharmacotherapeutic molecules delivery, diagnostics, as well as imaging [2, 24–26, 28].

In this chapter, we focus on fluorescent nanomaterials which are used as a probe for detection, biomedical imaging, and diagnosis. Their properties and applications in different species will be discussed; their toxicity to the test organisms will be evaluated. Mammalian cell lines, rotifers, Gram-negative bacteria, Gram-positive bacteria, and archaea are the choices of our topic. The 1-(4,5-dimethylthiazol-2-yl)-3,5-diphenylformazan (MTT) and sulforhodamine B (SRB) assays are indicative of cytotoxicity.

1.1 Fluorescent nanoparticles

Bioimaging is increasingly popular and important because of its noninvasive, dynamic, and real-time properties. Fluorescent molecules play a pivotal role in optical imaging of life science research and biomedical applications. Fluorescent probes can be used to detect RNA and DNA; analyze proteins, hormones, and viral antigens; and identify organelles, specific proteins, tissues, or tumors via antibody conjugation [38]. However, traditional fluorescent probes made by organic compounds have a lot of room for improvement such as sensitivity, fluorescent intensity, water-solubility, and photo-bleaching [38]. Another critical issue is their asymmetric excitation spectrum with a long tail leading to significant overlapping with other fluorescent dyes [38]. Various fluorescent nanomaterials have high quantum yield and permanent fluorescence with the potential to 1 day replacing the traditional probes [39]. Advantages of fluorescent nanoparticles include, but not limited to, tunable in both sizes and compositions, ultra-bright with narrow spectrum of emission but broad spectrum of excitation, resistant to chemical degradation and photobleaching for long-term observation [40, 41]. Semiconductor nanoparticles such as CdSe/ZnS quantum dots (QDs) are the best example in bioimaging, diagnostics, and therapeutic molecules delivery [42]. Additional fluorescent nanomaterials have been developed to serve as bioimaging probes. We describe some popular fluorescent nanoparticles as follows.

1.1.1 Cadmium-based QDs

Cadmium-based QDs are the first colloidal semiconductor which started the new era of nanotechnology in bioimaging [43]. Extreme brightness and sharp peak of emission wavelength are their signature properties [40–43]. High stability in the biological environment makes them applicable for long-term bioimaging [40–43]. The versatile surface modifications of their outer shells greatly expand their functionalities and biocompatibilities [40–43].

1.1.2 Indium-based QDs

Indium phosphate/zinc sulfate (InP/ZnS) QDs which belongs to the III-V groups of semiconductors replace the toxic cadmium core as a new generation of nanoparticles in biomedical imaging [44]. InP/ZnS QDs modified with PEG-containing negative charges enable them to interact with cationic peptides [44]. PEG modification also increase stability in the aqueous environment. These new complexes entered cells via endocytosis without toxicity [44].

1.1.3 Graphene quantum dots (GQDs)

Graphene is a two-dimensional sheet structure with single-atom thickness. GQDs have attracted plenty of attention due to their low toxicity, water solubility, high stability, stable emission spectrum, better surface grafting, and high electrical and thermal conductivity [45]. There are a variety of synthesis methods for GQDs that can control the shape, size, and yield [45].

1.1.4 Carbon dots (CDs)

The fluorescent property of carbon dots were found accidentally [39]. CDs are low cytotoxic, eco-friendly, and highly biocompatible. These features allow them to overtake cadmium-cored QDs in bioimaging [39]. Safety studies have shown that they neither enter nucleus and damage chromosomes nor accumulate in mice bodies [39].

1.1.5 Zirconium porphyrinic metal-organic framework nanoparticles (ZrMOF)

Metal–organic frameworks (MOF) combined the inorganic and organic materials and the distinguished features are tunable pore size, high surface areas, and alterable internal surface properties. MOFs do not emit fluorescence, but can be loaded with organic dyes for biomedical sensing. Quenching and therefore became the target-induced sensor [46]. ZrMOF nanoparticles can be unquenched in the target site, thereby facilitating the effect of photodynamic cancer therapies [46].

1.1.6 Soft fluorescent nanomaterials

Soft fluorescent nanomaterials include dye-doped polymer, semiconducting polymer, organic-complex nanoparticles, micelles, and nanogels [47]. This group of fluorescent nanomaterials is a complicated convention. For example, semiconductor QDs are composed of V-III or VI-II groups of elements in their cores and shells, but they can also be functionized by organic materials via chemical linkage on the surface [41, 44]. Modified QDs are classified as soft fluorescent nanomaterials while FDA approved molecules are used and these complexes are fabricated under a mild condition [47].

1.2 Biocompatible enhancer for nanoparticle delivery

Most organic and inorganic nanomaterials including the nanoparticles of the six categories described above are either hydrophobic or water-insoluble, which means they are difficult to be delivered into cells or organisms. Lipid-derived or peptide-based modifications of shells on the surfaces of nanoparticles enhance the biocompatibility and increase the efficiency of cellular uptake [48, 49]. The routes of cellular deliveries in these nanoparticles functionized by lipid-derived or peptide-based linking chains are related to endocytosis [50, 51]. Antibody-functionalized nanoparticles are advantageous in targeting [51]; however, lysosomal degradation of the delivered drugs or bioactive molecules was a major concern. Several strategies have been developed to circumvent this issue. Among them are cell-penetrating peptides (CPPs) [52–54].

CPPs (a.k.a., PTDs), are proteins that possess the ability to penetrate cell membranes. In recent decades, they have attracted immense popularity in delivering genes, bioactive macromolecules, and drugs due to their effect intracellular translocation. The first CPPs were identified in the human immunodeficiency virus type I (HIV-1) transcriptional activator Tat which consists of 11-amino acids (YGRKKRRQRRR) [55, 56]. Later, various natural CPPs were found in many organisms and numerous synthetic CPPs were designed [57]. To date, there are more than 1700 CPPs in the databank [58]. CPPs can be classified into three categories: cationic, amphipathic, and hydrophobic [57]. Modifications such as cyclization, branches, D-form alteration, and non-primary amino-acid utilization have been performed to improve their transduction efficiency [59–62]. The mechanisms by which CPPs internalized have been vigorously studied. Numerous studies suggest that depending on the physicochemical properties and secondary structures of CPPs, energy-dependent endocytosis or direct membrane translocation is employed [52, 57]. Four major endocytoses are: clathrin-dependent, caveolae-dependent, clathrin and caveolae-independent, and macropinocytosis [57]. Entry of molecules via these four pathways is trapped in lysosomes and their bioactive and therapeutic characteristics would be lost in the low pH environment [63]. CPP-mediated direct membrane location is an option for delivering drugs and other bioactive molecules [52, 53, 64].

CPPs can interact with cargoes in a covalent, noncovalent, or covalent and noncovalent protein transduction (CNPT) manner. Cargoes are various including proteins, DNA, siRNA, and semiconductors QDs [44, 64–68]. CPP-mediated cellular uptake can be found from prokaryotic to eukaryotic organisms including mammalian cells, insect cells, aquatic microorganism, yeasts, plant tissue, mice dermis, Gram-negative and Gram-positive bacteria, and archaea [53, 65, 69–73]. Uptake mechanisms vary depending upon peptide sequences. IR9 consists of the INF7 fusion peptide and nona-arginine CPP. IR9 without taking any cargoes penetrated cell membranes via macropinocytosis. However, when IR9 was mixed with DNAs or QDs, classical endocytosis was utilized [74]. HR9 contains nona-arginine in the center, flanks with five histidines on either side, and caps with a cysteine in both ends. Out study showed that HR9-mediated cellular entry involves direct membrane translocation [53, 75].

PR9 which consists of nona-arginine and a penetration accelerating peptide sequence has been used to deliver ODs. The complexes entered cells by classical endocytosis [54]. Subsequently, the PR9/QD complexes escaped from lysosomes and entered nucleus [54]. The fluorescent quantum yield and complexes properties were unaltered indicating that CPP/QD complexes were suitable for long-term intracellular imaging and tracking [54, 77, 78].

In the following sections, we discuss biocompatibility of CPP-mediated delivery in various systems including mammalian cells, rotifers, Gram-negative bacteria, Gram-positive bacteria, and archaea.

2. Cell viability in mammalian cells

In our studies, human bronchoalveolar carcinoma A549 cells were used as a model cell line to investigate CPP-mediated uptake of inorganic fluorescent nanoparticles which are CdSe/ZnS QD with green fluorescence, InP/ZnS emitted green fluorescence, and CdSe/ZnS QD with red fluorescence. Synthetic nona-arginine (named SR9) CPP were premixed with these three QDs respectively and incubated

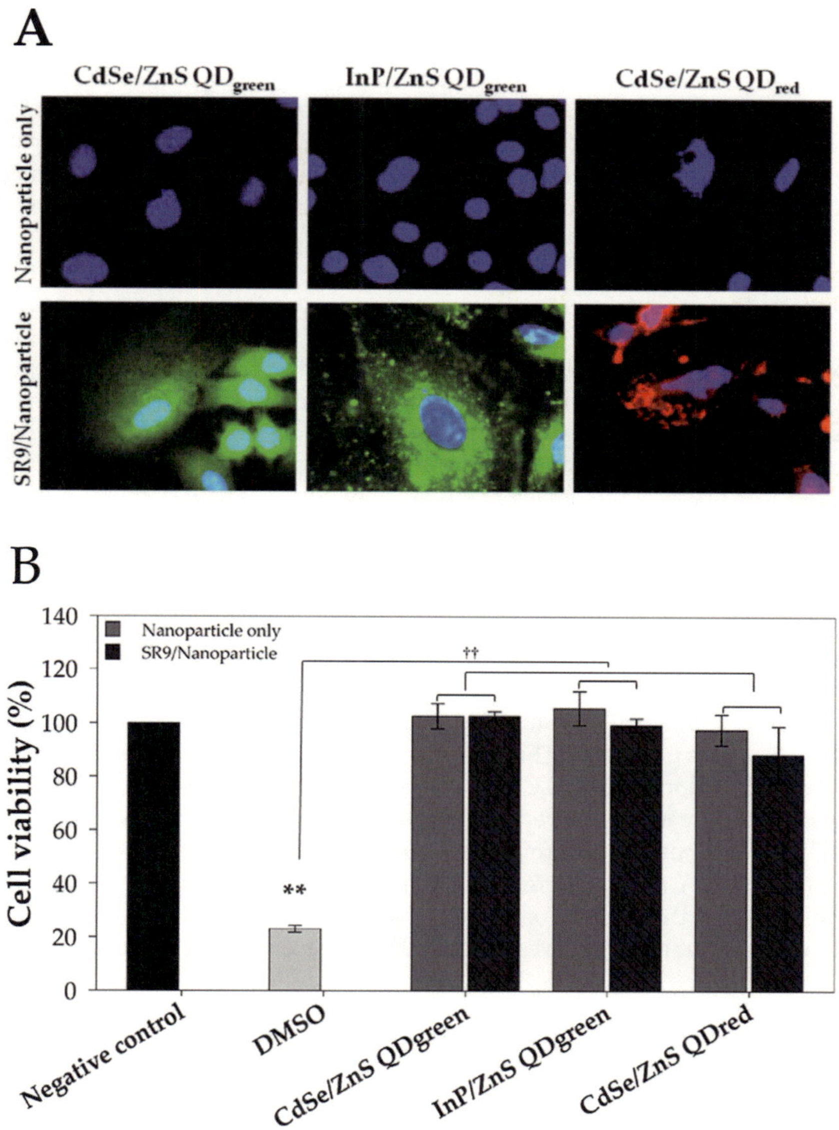

Figure 1.
*Semiconductor nanoparticles treatments in the mammalian A549 cell line. (A) Penetrations of various semiconductor nanoparticles and CPP/nanoparticles complexes in mammalian cells. A549 cells were treated with green fluorescent CdSe/ZnS QD, green fluorescent InP/ZnS QD, red fluorescent CdSe/ZnS QD, SR9/CdSe/ZnS QD_{green} complexes, SR9/InP/ZnS QD_{green} complexes, and SR9/CdSe/ZnS QD_{red} for 1 h at 37°C, respectively. Protein transductions were recorded using a BD pathway 435 system. Green and red fluorescence revealed the distribution of nanoparticles, and blue fluorescence indicated the nuclei. Images were taken at a magnification of 600×. (B) Cell viabilities in A549 cells treated with either nanoparticles alone or CPP/nanoparticles complexes. Cells were treated as previous description shown in (A) and the SRB assay was performed for cytotoxic analysis. Cells without any treatments and treated with 100% DMSO were served as the negative and positive groups, respectively. Histogram of cell viability was represented by mean ± SD from three independent experiments in each treatment group. Significant differences at $P < 0.01$ (**,††) are indicated [44, 53, 78].*

with A549 cells for 1 h followed by Hoechst 33342 nuclear staining. Internalizations of QDs and SR9/QD complexes were determined using confocal microscopy. Rare green fluorescence emitted from CdSe/ZnS QD_{green} and InP/ZnS QD_{green} as well as red fluorescence from CdSe/ZnS QD_{red} were observed while cells were treated with these semiconductors alone (**Figure 1A**). However, strong green and red fluorescence were observed in the groups of SR9-mediated QDs delivery, which meant SR9 facilitated the internalizations of nanoparticles (**Figure 1A**).

To understand toxicity of these CPP-QDs complexes, the SRB assay was conducted for viability analysis. A549 cells were treated with the materials for 24 h and then stained with SRB. Cells without any treatment or with 100% DMSO served as a negative control or a positive control, respectively. Cell viability of CdSe/ZnS QD_{green}, InP/ZnS QD_{green}, CdSe/ZnS QD_{red}, and SR9-modified CdSe/ZnS QD_{green}, InP/ZnS QD_{green}, and CdSe/ZnS QD_{red} complexes did not differ from the negative control (**Figure 1B**). Collectively, semiconductor fluorescent nanoparticles and their CPP-modified complexes did not reduce cell viability.

3. Survival rate in rotifers

Rotifers are non-arthropoda, metazoan aquatic invertebrates with a completed digestive systems. They form the basis of the microzooplankton community in the plankton food web and link the energy flow to higher organisms. Recently, a growing number of studies considers rotifers as an indicator of marine pollution and toxicity of plastic nanoparticles [79–81], as well as a model species for pharmaceutical and toxicological studies [80, 82]. To test the toxicity and uptake efficiency of fluorescent nanoparticles, *Brachionus calyciflorus* were treated with CdSe/ZnS QD_{red} and IR9-FITC mixed CdSe/ZnS QD_{red} complexes respectively (**Figure 2**). Low red fluorescent intensity was detected in rotifers which meant CdSe/ZnS QD_{red} were difficult to enter rotifers without any help. Contrarily, CdSe/ZnS QD_{red} enter rotifers easily by forming complexes with IR9-FITC (**Figure 2A**).

To investigate potential cytotoxicity of CPP-associated quantum dots on rotifers, the MTT assay was performed (**Figure 1B**). *Brachionus calyciflorus* were treated with CdSe/ZnS QD_{red} alone, IR9-FITC alone, and IR9-FITC/QD_{red} complexes for 24 h. Rotifers without treatment served as a negative control, while rotifers treated with 100% DMSO as a positive control. Hypotoxicity was observed in the QD_{red}, IR9-FITC, and IR9-FITC/QD_{red} complexes groups (**Figure 2B**). In contrast, DMSO significantly reduced the survival of rotifers (**Figure 2B**). Collectively, CPP-mediated cellular entry of quantum dots resulted in relatively harmless in rotifers.

4. Hypotoxicity shown in prokaryotic organisms

Microorganisms are regarded as a vital member in the ecosystem as they play an important role in the natural recycling, elements and energy transforming, and environmental balancing of living materials [83, 84]. Disruption of microorganisms cause reduction of microbial diversity and obliquely influence our natural world [85]. Prokaryotic organisms are major microorganisms; the prokaryotic domain include bacteria and archaea [70, 86, 87]. Here, *Arthrobacter ilicis* D50–1 (Gram-positive bacteria), *Escherichia coli* DH5α (Gram-negative bacteria), and *Thermus aquaticus* (archaea) were studied for protein transduction and cytotoxicity. They were treated with Cd-core green semiconductor nanoparticles; hardly any green fluorescence detected (**Figure 3A**). Bright green fluorescence was observed in the SR9-mediated uptake of CdSe/ZnS QD_{green} in all three organisms (**Figure 3B**).

Toxicological studies of nanomaterials on prokaryotic organisms are important. Bactericidal nanomaterials can affect nonpathogenic bacteria leading to imbalance of a microbiome community and, to the greatest extent, ecological disasters [88]. The toxicity of CdSe/ZnS QD_{green} and SR9-modified QD_{green} complexes were studied using archaea, Gram-positive bacteria, and Gram-negative bacteria (**Figure 3B**). Organisms were treated for 1 h at room temperature. The MTT assay showed no reduction of viability

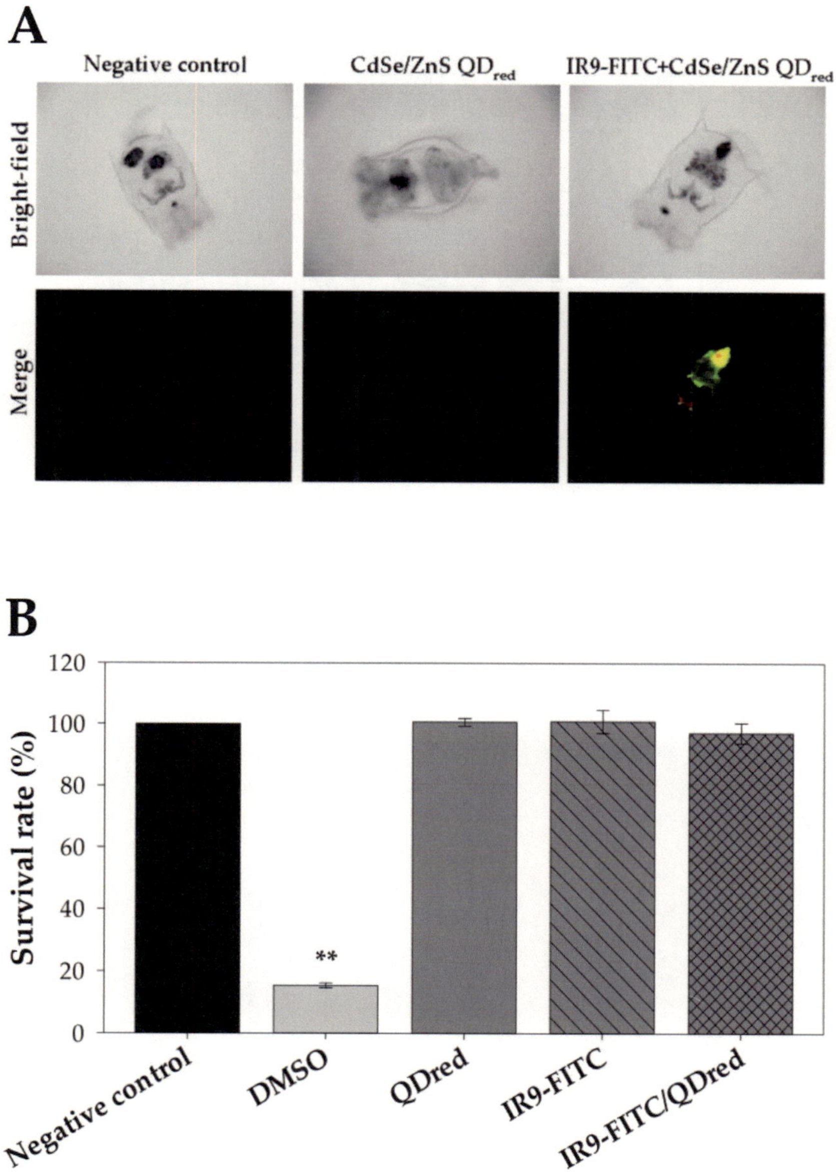

Figure 2.
Semiconductor nanoparticles treatments in the rotifer (Brachionus calyciflorus). (A) Protein transduction in rotifer treated with semiconductor nanoparticles CdSe/ZnS QD_{red} and IR9-FITC carpeted QD_{red} complexes. Rotifers were treated with QD_{red} alone or IR9-FITC/QD_{red} complexes for 1 h at 25–28°C. Green fluorescence referred to IR9-FITC and red fluorescence indicated QD_{red}. Merged fluorescent images and bright-field images were recorded at a magnification of 200× using a BD pathway 435 system. (B) Histogram of rotifer survival rate. Rotifers were treated with CdSe/ZnS QD_{red} alone, IR9-FITC alone, and IR9-FITC/QD_{red} complexes, respectively and the survival rate was analyzed by MTT assay. Rotifers treated with water and 100% DMSO were served as negative and positive controls, respectively. Each treatment group was compared with the negative control. Significant differences at $P < 0.01$ (**) were indicated. Data were presented as mean ± standard deviation from three independent experiments in each treatment groups [74].

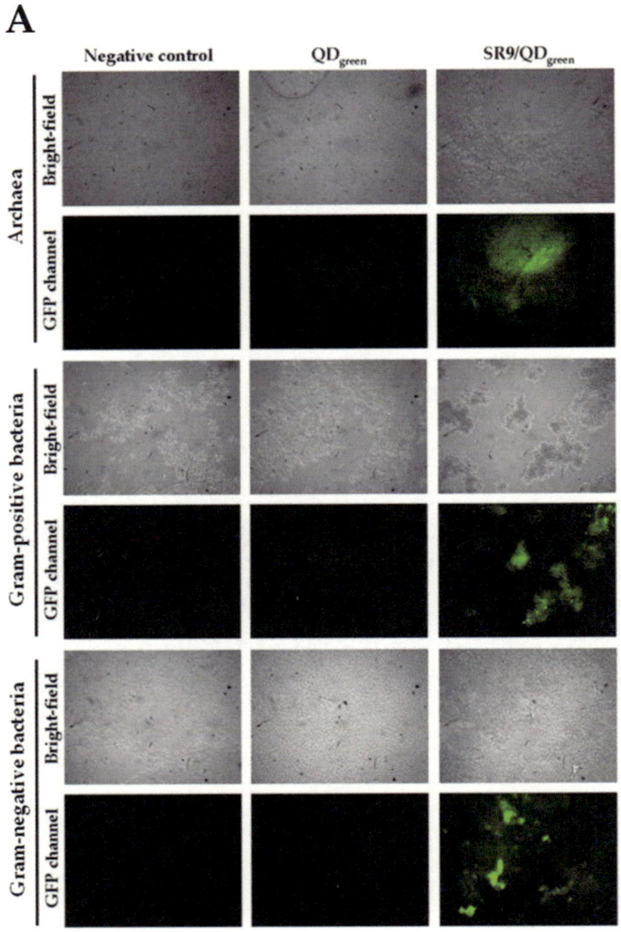

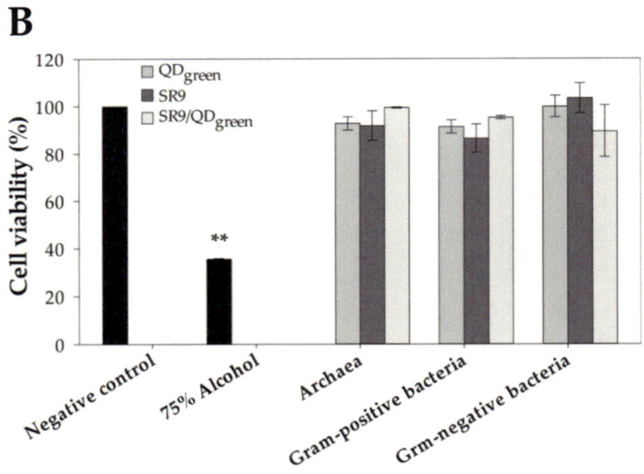

Figure 3.

Treatments of CdSe cored QD with green fluorescence in three types of prokaryocytes. (A) Fluorescent microscopy of Thermus aquaticus (archaea), Arthrobacter ilicis D50-1 (Gram-positive bacteria), and Escherichia coli DH5α (Gram-negative bacteria) treated with QD_{green} alone or SR9/QD_{green} complexes. Three prokaryocytes were incubated with QD_{green} alone or SR9/QD_{green} complexes for 1 h at room temperature. Protein transductions were detected in GFP channel and cell morphologies were observed in bright-field with an AE31 fluorescent microscope. (B) Cytotoxicity of nanoparticles in archaea, Gram-positive bacteria, and Gram-negative bacteria. Archaea, D50–1, and DH5α were treated with QD_{green}, SR9, and SR9/QD_{green} complexes. Cells treated with their specific media and 75% alcohol as negative control and positive control, respectively. Significant differences from negative control at $P < 0.01$ (**) were indicated. Data were presented as mean ± standard deviation from three independent experiments [89].

in the groups of QD_{green}, SR9, and SR9/QD_{green} complexes (**Figure 3B**). This result indicated that Cd-core nanoparticles did not cause lethal effect to prokaryotic organisms. We reasonably ratiocinated that fluorescent nanoparticles applied in bioimaging and biotechnologies might not provoke natural imbalance and environmental problems.

5. Conclusion

We discussed applications and safety issues of various fluorescent nanoparticles. The cellular entry of particles of interest can be facilitated by CPPs. The particles did not produce lethal effects in mammalian cells, rotifers, archaea, Gram-positive bacteria, and Gram-negative bacteria. The outcome from assessing nanoparticle safety in mammalian cells suggests their potential medical applications. Hypotoxicity in rotifers and prokaryotes infers their environmental safety and eco-friendliness. In summary, these fluorescent nanoparticles and their CPP-modified complexes can be potent tools in various biological, environmental, and medical applications in the future.

Author details

Betty Revon Liu[1*], Yue-Wern Huang[2] and Han-Jung Lee[3]

1 Department of Laboratory Medicine and Biotechnology, Tzu Chi University, Hualien, Taiwan

2 Department of Biological Sciences, Missouri University of Science and Technology, Rolla, MO, USA

3 Department of Natural Resources and Environmental Studies, National Dong Hwa University, Hualien, Taiwan

*Address all correspondence to: brliu7447@gms.tcu.edu.tw

References

[1] Kagan CR, Fernandez LE, Gogotsi Y, Hammond PT, Hersam MC, et al. Nano day: Celebrating the next decade of nanoscience and nanotechnology. ACS Nano. 2016;**10**:9093-9103

[2] Salata OV. Applications of nanoparticles in biology and medicine. Journal of Nanobiotechnology. 2004;**2**:3-3

[3] Ostadhossein F, Pan D. Functional carbon nanodots for multiscale imaging and therapy. Wiley Interdisciplinary Reviews. Nanomedicine and Nanobiotechnology. 2017:9

[4] Lu X, Jia W, Chai H, Hu J, Wang S, et al. Solid-state chemical fabrication of one-dimensional mesoporous β-nickel molybdate nanorods as remarkable electrode material for supercapacitors. Journal of Colloid and Interface Science. 2019;**534**:322-331

[5] Patra S, Roy E, Tiwari A, Madhuri R, Sharma PK. 2-Dimensional graphene as a route for emergence of additional dimension nanomaterials. Biosensors and Bioelectronics. 2017;**89**:8-27

[6] Chimene D, Alge DL, Gaharwar AK. Two-dimensional nanomaterials for biomedical applications: Emerging trends and future prospects. Advanced Materials. 2015;**27**:7261-7284

[7] Ray PC. Size and shape dependent second order nonlinear optical properties of nanomaterials and their application in biological and chemical sensing. Chemical Reviews. 2010;**110**:5332-5365

[8] Usman MS, Zowalaty MEE, Shameli K, Zainuddin N, Salama M, et al. Synthesis, characterization, and antimicrobial properties of copper nanoparticles. International Journal of Nanomedicine. 2013;**8**:4467-4479

[9] Ganju A, Khan S, Hafeez BB, Behrman SW, Yallapu MM, et al. miRNA nanotherapeutics for cancer. Drug Discovery Today. 2017;**22**:424-432

[10] Lalwani G, Patel SC, Sitharaman B. Two- and three-dimensional all-carbon nanomaterial assemblies for tissue engineering and regenerative medicine. Annals of Biomedical Engineering. 2016;**44**:2020-2035

[11] Jeevanandam J, Barhoum A, Chan YS, Dufresne A, Danquah MK. Review on nanoparticles and nanostructured materials: History, sources, toxicity and regulations. Beilstein Journal of Nanotechnology. 2018;**9**:1050-1074

[12] Cha C, Shin SR, Annabi N, Dokmeci MR, Khademhosseini A. Carbon-based nanomaterials: Multifunctional materials for biomedical engineering. ACS Nano. 2013;**7**:2891-2897

[13] Golbamaki A, Golbamaki N, Sizochenko N, Rasulev B, Leszczynski J, et al. Genotoxicity induced by metal oxide nanoparticles: A weight of evidence study and effect of particle surface and electronic properties. Nanotoxicology. 2018:1-17

[14] Wu L, Zhang F, Wei Z, Li X, Zhao H, et al. Magnetic delivery of Fe_3O_4@ polydopamine nanoparticle-loaded natural killer cells suggest a promising anticancer treatment. Biomaterials Science. 2018;**6**:2714-2725

[15] Radhakrishnan VS, Reddy Mudiam MK, Kumar M, Dwivedi SP, Singh SP, et al. Silver nanoparticles induced alterations in multiple cellular targets, which are critical for drug susceptibilities and pathogenicity in fungal pathogen (*Candida albicans*). International Journal of Nanomedicine. 2018;**13**:2647-2663

[16] Ana CS, Carla ML, José MSL, Maria HA. Delivery systems for biopharmaceuticals. Part II: Liposomes, micelles, microemulsions and dendrimers. Current Pharmaceutical Biotechnology. 2015;**16**:955-965

[17] Khan I, Saeed K, Khan I. Nanoparticles: Properties, applications and toxicities. Arabian Journal of Chemistry. 2017

[18] Bhatia S. Nanoparticles Types, Classification, Characterization, Fabrication Methods and Drug Delivery Applications. Natural Polymer Drug Delivery Systems. Cham: Springer; 2016. pp. 33-93

[19] McDevitt MR, Chattopadhyay D, Kappel BJ, Jaggi JS, Schiffman SR, et al. Tumor targeting with antibody-functionalized, radiolabeled carbon nanotubes. Journal of Nuclear Medicine. 2007;**48**:1180-1189

[20] Zhang Z, Yang X, Zhang Y, Zeng B, Wang S, et al. Delivery of telomerase reverse transcriptase small interfering RNA in complex with positively charged single-walled carbon nanotubes suppresses tumor growth. Clinical Cancer Research. 2006;**12**:4933-4939

[21] Ji H, Yang Z, Jiang W, Geng C, Gong M, et al. Antiviral activity of nano carbon fullerene lipidosome against influenza virus in vitro. Journal of Huazhong University of Science and Technology. Medical Sciences. 2008;**28**:243-246

[22] Levin A, Michaels TCT, Mason TO, Muller T, Adler-Abramovich L, et al. Self-assembly-mediated release of peptide nanoparticles through jets across microdroplet interfaces. ACS Applied Materials & Interfaces. 2018;**10**:27578-27583

[23] Pan B, Cui D, Sheng Y, Ozkan C, Gao F, et al. Dendrimer-modified magnetic nanoparticles enhance efficiency of gene delivery system. Cancer Research. 2007;**67**:8156-8163

[24] Tomalia DA, Reyna LA, Svenson S. Dendrimers as multi-purpose nanodevices for oncology drug delivery and diagnostic imaging. Biochemical Society Transactions. 2007;**35**:61-67

[25] Elzoghby AO, Abd-Elwakil MM, Abd-Elsalam K, Elsayed MT, Hashem Y, et al. Natural polymeric nanoparticles for brain-targeting: Implications on drug and gene delivery. Current Pharmaceutical Design. 2016;**22**:3305-3323

[26] Amiot CL, Xu S, Liang S, Pan L, Zhao JX. Near-infrared fluorescent materials for sensing of biological targets. Sensors (Basel). 2008;**8**:3082-3105

[27] Moore A, Weissleder R, Bogdanov A Jr. Uptake of dextran-coated monocrystalline iron oxides in tumor cells and macrophages. Journal of Magnetic Resonance Imaging. 1997;**7**:1140-1145

[28] Gao X, Cui Y, Levenson RM, Chung LW, Nie S. In vivo cancer targeting and imaging with semiconductor quantum dots. Nature Biotechnology. 2004;**22**:969-976

[29] Tegos GP, Demidova TN, Arcila-Lopez D, Lee H, Wharton T, et al. Cationic fullerenes are effective and selective antimicrobial photosensitizers. Chemistry & Biology. 2005;**12**:1127-1135

[30] Markovic Z, Trajkovic V. Biomedical potential of the reactive oxygen species generation and quenching by fullerenes (C60). Biomaterials. 2008;**29**:3561-3573

[31] Freitas RA Jr. Pharmacytes: An ideal vehicle for targeted drug delivery. Journal of Nanoscience and Nanotechnology. 2006;**6**:2769-2775

[32] Cai X, Jia H, Liu Z, Hou B, Luo C, et al. Polyhydroxylated fullerene derivative C(60)(OH)(24) prevents mitochondrial dysfunction and oxidative damage in an MPP(+) -induced cellular model of Parkinson's disease. Journal of Neuroscience Research. 2008;**86**:3622-3634

[33] Iverson N, Plourde N, Chnari E, Nackman GB, Moghe PV. Convergence of nanotechnology and cardiovascular medicine: Progress and emerging prospects. Bio Drugs. 2008;**22**:1-10

[34] Heerema SJ, Dekker C. Graphene nanodevices for DNA sequencing. Nature Nanotechnology. 2016;**11**:127-136

[35] Li BR, Chen CC, Kumar UR, Chen YT. Advances in nanowire transistors for biological analysis and cellular investigation. Analyst. 2014;**139**:1589-1608

[36] Zhang L, Liu Z, Yang K, Kong C, Liu C, et al. Tumor progression of non-small cell lung cancer controlled by albumin and micellar nanoparticles of itraconazole, a multitarget angiogenesis inhibitor. Molecular Pharmaceutics. 2017;**14**:4705-4713

[37] Kim GH, Won JE, Byeon Y, Kim MG, Wi TI, et al. Selective delivery of PLXDC1 small interfering RNA to endothelial cells for anti-angiogenesis tumor therapy using CD44-targeted chitosan nanoparticles for epithelial ovarian cancer. Drug Delivery. 2018;**25**:1394-1402

[38] Fei X, Gu Y. Progress in modifications and applications of fluorescent dye probe. Progress in Natural Science. 2009;**19**:1-7

[39] Zuo J, Jiang T, Zhao X, Xiong X, Xiao S, et al. Preparation and application of fluorescent carbon dots. Journal of Nanomaterials. 2015, 2015:13

[40] Michalet X, Pinaud FF, Bentolila LA, Tsay JM, Doose S, et al. Quantum dots for live cells, in vivo imaging, and diagnostics. Science. 2005;**307**:538-544

[41] Chen F, Gerion D. Fluorescent CdSe/ZnS nanocrystal–peptide conjugates for long-term, nontoxic imaging and nuclear targeting in living cells. Nano Letters. 2004;**4**:1827-1832

[42] Bilan R, Nabiev I, Sukhanova A. Quantum dot-based nanotools for bioimaging, diagnostics, and drug delivery. Chembiochem. 2016;**17**:2103-2114

[43] Rizvi SB, Ghaderi S, Keshtgar M, Seifalian AM. Semiconductor quantum dots as fluorescent probes for in vitro and in vivo bio-molecular and cellular imaging. Nano Reviews. 2010;**1**. DOI: 10.3402/nano.v3401i3400.5161

[44] Liu BR, Winiarz JG, Moon JS, Lo SY, Huang YW, et al. Synthesis, characterization and applications of carboxylated and polyethylene-glycolated bifunctionalized InP/ZnS quantum dots in cellular internalization mediated by cell-penetrating peptides. Colloids and Surfaces. B, Biointerfaces. 2013;**111**:162-170

[45] Liu Q, Zhang J, He H, Huang G, Xing B, et al. Green preparation of high yield fluorescent graphene quantum dots from coal-tar-pitch by mild oxidation. Nanomaterials. 2018;**8**:844

[46] Liu Y, Hou W, Xia L, Cui C, Wan S, et al. ZrMOF nanoparticles as quenchers to conjugate DNA aptamers for target-induced bioimaging and photodynamic therapy. Chemical Science. 2018;**9**:7505-7509

[47] Peng H-S, Chiu DT. Soft fluorescent nanomaterials for biological and biomedical imaging. Chemical Society Reviews. 2015;**44**:4699-4722

[48] Ranalli A, Santi M, Capriotti L, Voliani V, Porciani D, et al. Peptide-based stealth nanoparticles for targeted and pH-triggered delivery. Bioconjugate Chemistry. 2017;**28**:627-635

[49] Andonova V, Peneva P. Characterization methods for solid lipid nanoparticles (SLN) and nanostructured lipid carriers (NLC). Current Pharmaceutical Design. 2017

[50] Oh N, Park J-H. Endocytosis and exocytosis of nanoparticles in mammalian cells. International Journal of Nanomedicine. 2014;**9**(Suppl 1):51-63

[51] Wu FL, Zhang J, Li W, Bian BX, Hong YD, et al. Enhanced antiproliferative activity of antibody-functionalized polymeric nanoparticles for targeted delivery of anti-miR-21 to HER2 positive gastric cancer. Oncotarget. 2017;**8**:67189-67202

[52] Liu BR, Huang YW, Korivi M, Lo SY, Aronstam RS, et al. The primary mechanism of cellular internalization for a short cell-penetrating peptide as a nano-scale delivery system. Current Pharmaceutical Biotechnology. 2017;**18**:569-584

[53] Liu BR, Huang YW, Winiarz JG, Chiang HJ, Lee HJ. Intracellular delivery of quantum dots mediated by a histidine- and arginine-rich HR9 cell-penetrating peptide through the direct membrane translocation mechanism. Biomaterials. 2011;**32**:3520-3537

[54] Liu BR, Lo SY, Liu CC, Chyan CL, Huang YW, et al. Endocytic trafficking of nanoparticles delivered by cell-penetrating peptides comprised of nona-arginine and a penetration accelerating sequence. PLoS One. 2013;**8**:e67100

[55] Frankel AD, Pabo CO. Cellular uptake of the tat protein from human immunodeficiency virus. Cell. 1988;**55**:1189-1193

[56] Schwarze SR, Ho A, Vocero-Akbani A, Dowdy SF. In vivo protein transduction: Delivery of a biologically active protein into the mouse. Science. 1999;**285**:1569-1572

[57] Guidotti G, Brambilla L, Rossi D. Cell-penetrating peptides: From basic research to clinics. Trends in Pharmacological Sciences. 2017;**38**:406-424

[58] Gautam A, Singh H, Tyagi A, Chaudhary K, Kumar R, et al. CPPsite: A curated database of cell penetrating peptides. Database: The Journal of Biological Databases and Curation. 2012, bas015

[59] Qian Z, Liu T, Liu YY, Briesewitz R, Barrios AM, et al. Efficient delivery of cyclic peptides into mammalian cells with short sequence motifs. ACS Chemical Biology. 2013;**8**:423-431

[60] Ezzat K, Andaloussi SE, Zaghloul EM, Lehto T, Lindberg S, et al. PepFect 14, a novel cell-penetrating peptide for oligonucleotide delivery in solution and as solid formulation. Nucleic Acids Research. 2011;**39**:5284-5298

[61] Farrera-Sinfreu J, Giralt E, Castel S, Albericio F, Royo M. Cell-penetrating cis-gamma-amino-l-proline-derived peptides. Journal of the American Chemical Society. 2005;**127**:9459-9468

[62] Angeles-Boza AM, Erazo-Oliveras A, Lee YJ, Pellois JP. Generation of endosomolytic reagents by branching of cell-penetrating peptides: Tools for the delivery of bioactive compounds to live cells in cis or trans. Bioconjugate Chemistry. 2010;**21**:2164-2167

[63] Hogset A, Prasmickaite L, Selbo PK, Hellum M, Engesaeter BO, et al. Photochemical internalisation in drug and gene delivery. Advanced Drug Delivery Reviews. 2004;**56**:95-115

[64] Liu BR, Huang YW, Aronstam RS, Lee HJ. Identification of a short cell-penetrating peptide from bovine Lactoferricin for intracellular delivery of DNA in human A549 cells. PLoS One. 2016;**11**:e0150439

[65] Chang M, Chou JC, Chen CP, Liu BR, Lee HJ. Noncovalent protein transduction in plant cells by macropinocytosis. The New Phytologist. 2007;**174**:46-56

[66] Chang M, Chou JC, Lee HJ. Cellular internalization of fluorescent proteins via arginine-rich intracellular delivery peptide in plant cells. Plant & Cell Physiology. 2005;**46**:482-488

[67] Hu JW, Liu BR, Wu CY, Lu SW, Lee HJ. Protein transport in human cells mediated by covalently and noncovalently conjugated arginine-rich intracellular delivery peptides. Peptides. 2009;**30**:1669-1678

[68] Wang YH, Hou YW, Lee HJ. An intracellular delivery method for siRNA by an arginine-rich peptide. Journal of Biochemical and Biophysical Methods. 2007;**70**:579-586

[69] Hou YW, Chan MH, Hsu HR, Liu BR, Chen CP, et al. Transdermal delivery of proteins mediated by non-covalently associated arginine-rich intracellular delivery peptides. Experimental Dermatology. 2007;**16**:999-1006

[70] Liu BR, Chou JC, Lee HJ. Cell membrane diversity in noncovalent protein transduction. The Journal of Membrane Biology. 2008;**222**:1-15

[71] Chen YJ, Liu BR, Dai YH, Lee CY, Chan MH, et al. A gene delivery system for insect cells mediated by arginine-rich cell-penetrating peptides. Gene. 2012;**493**:201-210

[72] Dai YH, Liu BR, Chiang HJ, Lee HJ. Gene transport and expression by arginine-rich cell-penetrating peptides in paramecium. Gene. 2011;**489**:89-97

[73] Liu BR, Liou JS, Chen YJ, Huang YW, Lee HJ. Delivery of nucleic acids, proteins, and nanoparticles by arginine-rich cell-penetrating peptides in rotifers. Marine Biotechnology (New York, N.Y.). 2013;**15**:584-595

[74] Liu BR, Liou JS, Huang YW, Aronstam RS, Lee HJ. Intracellular delivery of nanoparticles and DNAs by IR9 cell-penetrating peptides. PLoS One. 2013;**8**:e64205

[75] Liu BR, Lin MD, Chiang HJ, Lee HJ. Arginine-rich cell-penetrating peptides deliver gene into living human cells. Gene. 2012;**505**:37-45

[76] Sangtani A, Petryayeva E, Wu M, Susumu K, Oh E, et al. Intracellularly actuated quantum dot-peptide-doxorubicin nanobioconjugates for controlled drug delivery via the endocytic pathway. Bioconjugate Chemistry. 2018;**29**:136-148

[77] Huang YW, Lee HJ, Liu BR, Chiang HJ, Wu CH. Cellular internalization of quantum dots. Methods in Molecular Biology. 2013;**991**:249-259

[78] Liu BR, Chen HH, Chan MH, Huang YW, Aronstam RS, et al. Three arginine-rich cell-penetrating peptides facilitate cellular internalization of red-emitting quantum dots. Journal of Nanoscience and Nanotechnology. 2015;**15**:2067-2078

[79] Anjusha A, Jyothibabu R, Jagadeesan L, Arunpandi N. Role of rotifers in microzooplankton community in a large monsoonal estuary (Cochin backwaters) along the west coast of India. Environmental Monitoring and Assessment. 2018;**190**:295

[80] Jeong CB, Kang HM, Lee YH, Kim MS, Lee JS, et al. Nanoplastic ingestion enhances toxicity of persistent

organic pollutants (POPs) in the monogonont rotifer *Brachionus koreanus* via multixenobiotic resistance (MXR) disruption. Environmental Science & Technology. 2018;**52**:11411-11418

[81] Manfra L, Rotini A, Bergami E, Grassi G, Faleri C, et al. Comparative ecotoxicity of polystyrene nanoparticles in natural seawater and reconstituted seawater using the rotifer *Brachionus plicatilis*. Ecotoxicology and Environmental Safety. 2017;**145**:557-563

[82] Dahms HU, Hagiwara A, Lee JS. Ecotoxicology, ecophysiology, and mechanistic studies with rotifers. Aquatic Toxicology. 2011;**101**:1-12

[83] Dalcin Martins P, Danczak RE, Roux S, Frank J, Borton MA, et al. Viral and metabolic controls on high rates of microbial sulfur and carbon cycling in wetland ecosystems. Microbiome. 2018;**6**:138

[84] Barberan A, Casamayor EO, Fierer N. The microbial contribution to macroecology. Frontiers in Microbiology. 2014;**5**:203

[85] Gibbons SM, Gilbert JA. Microbial diversity—Exploration of natural ecosystems and microbiomes. Current Opinion in Genetics & Development. 2015;**35**:66-72

[86] Guerrero R. Bergey's manuals and the classification of prokaryotes. International Microbiology. 2001;**4**:103-109

[87] Woese CR, Fox GE. Phylogenetic structure of the prokaryotic domain: The primary kingdoms. Proceedings of the National Academy of Sciences of the United States of America. 1977;**74**:5088-5090

[88] Wilson N. Nanoparticles: Environmental problems or problem solvers? Bioscience. 2018;**68**:241-246

[89] Liu BR, Huang YW, Aronstam RS, Lee HJ. Comparative mechanisms of protein transduction mediated by cell-penetrating peptides in prokaryotes. Journal of Membrane Biology. 2015;**248**:355-368

Chapter 5

rhBMP-2-Coated Acellular Dermal Graft for Chronic Rotator Cuff Healing: Translational Tendon Repair Research

Kwang-Il Lee, Ju-Woong Jang and Kwang-Won Lee

Abstract

A rotator cuff tear is a common shoulder injury in sports medicine. However, a rotator cuff repair still has the high failure rate (57%) in large torn (>8 cm^2) rotator cuff cases. One of the main reasons is failing at suture-tendon cause of continuous tensional and torsional stresses even after surgery, and thus, an ideal biologic augmentation to overcome large tears is an essential challenge. The ECM graft, the biological material can be useful for augment repair of large torn rotator cuff. Recombinant human bone morphogenetic protein 2 (rhBMP-2), which belongs to transforming growth factor-β superfamily, is well known as an osteoinductive growth factor. It plays an important role in the development of bone and cartilage. rhBMP-2 also facilitates chemotaxis in the host tissue. In this study, rhBMP-2-coated acellular dermal graft, which is isolated from human cadaveric donor, was transplanted in the rabbit with the chronic rotator cuff injury. The radiologic image, histomorphometric, histologic image analyses, and tensile test were performed to evaluate the effectiveness. The results showed the enhancement of increased host cell infiltration, new bone formation, and tensile mechanical property. The rhBMP-2-coated acellular dermal graft will be promising for chronic rotator cuff healing.

Keywords: rotator cuff, rhBMP-2, acellular dermis, tendon-bone healing, enthesis

1. Introduction

A rotator cuff plays an important role in shoulder movements and maintaining shoulder joint stability [1]. The rotator cuff is composed of supraspinatus muscle, subscapularis muscle, infraspinatus muscle, and teres minor muscle. The rotator cuff is connected between the scapula and the humeral bone head, forming a cuff at the shoulder joint. The rotator cuff tear, which is partial tear and full-thickness tear, mainly occurs in the supraspinatus tendon under the acromion [2]. A torn rotator cuff weakens shoulder's physical function. Almost 2 million people in the United States visit to see doctors cause of the painful rotator cuff at every year [3]. The rotator cuff repair requires reattachment of the rotator cuff tendon back to the humeral bone by using suture and fixation tools, which is non-biological technique [4, 5]. Rotator cuff repair has showed good long-term follow-up results so far. However, it still shows the high failure rate (57%) in the case of large tear (more than 8 cm^2) size [6]. The main reason of the pain even after rotator cuff repair surgery is that the shoulder has stiff due to the limited movement [7]. During the surgery, the rotator cuff tendons are sutured to the upper humeral bone. In the case of large tear size, it will be much harder to suture the shortened tendon back to the bone than the repair of the small tear size [8]. More tensile stress will be applied to the sutured gap between the tendon and bone.

The limitations for rotator cuff healing maybe overcome by biological augmentation approaches. A bio-scaffold should have high initial fixation strength, mechanical stability, and biological healing of the tendon-bone interface. And a growth factor should induce host cell infiltration and new tissue formation. Those characteristics are strongly necessary for biological augmentation approaches [9–12]. The dermal graft, which has been used for wound coverage, maybe applicable for large rotator cuff tears. The dermal graft is an acellular dermal extracellular matrix (ECM) scaffold intended for supporting and covering the soft tissue repair. There are various types of the acellular ECM scaffold such as small intestinal submucosa, urinary bladder basement membrane, pericardium, and dermis [13–16]. The only dermal graft has the most thickness (maximum 4 mm), and this will be the only ECM scaffold that can cover the thickness of supraspinatus (avg. 4.9 mm) or infraspinatus (avg. 4.2 mm) tendon [17]. Dermal graft has been used as a grafting material in plastic, dental, and orthopedic surgeries [18–20].

Various growth factors have been applied for enthesis regeneration studies [21–25]. Recombinant human bone morphogenetic protein 2 (rhBMP-2) is an osteoinductive protein that induces differentiation of mesenchymal stem cells into osteoblasts and chondrocytes. rhBMP-2 has shown the potential to improve tendon-bone biomechanical strength during rotator cuff healing [26, 27]. In our previous studies, we found that injectable rhBMP-2 hydrogel increased the biomechanical properties and new bone formation at the tendon-tibia intra articular bone tunnel [28].

The purpose of this translational study is to investigate whether the rhBMP-2-coated dermal graft can improve rotator cuff healing using a rabbit chronic rotator cuff injury model.

2. Materials and methods

2.1 Decellularization of human dermis and rhBMP-2 coating

Dermis was procured from the back of human cadaver and cut into pieces (1 × 2 cm). The tissue pieces were digested in an enzyme cocktail (total 500 mL) mixed with 0.25%(v/v) trypsin, 7.5%(w/v) collagenase A, and 37.5%(w/v) protease in a shaking incubator at 120 rpm, 37°C for 4 hours [29]. The tissues were then washed with the same volume of saline solution at 120 rpm, 4°C for 1 hour. This washing step was repeated for three times and finished with the final saline washing for 12 hours. And then, the washing solution was replaced with the same volume of distilled water and the tissues were cleansed with ultrasonic waves at 240 W for 5 minutes. This step was repeated in fresh saline solution. The processed tissues were lyophilized and sterilized with ethylene oxide gas.

Chinese hamster ovary cell line-derived rhBMP-2 was reconstituted and diluted with distilled water to 50 ug/mL before coating on the scaffold. rhBMP-2 was covered on the both side of the dermal graft and, then, lyophilized.

2.2 Surgery for making a chronic rotator cuff injury model

Adult New Zealand white rabbits (n = 42, male, 5-month old, 3.0 kg) were used for the chronic rotator cuff tear model. All the animal surgical protocols were approved by the Committee of Experimental Animal Sciences. To make the chronic injury model, the rabbits were anesthetized with ketamine, 40 mg/kg IM and xylazine, 5 mg/kg IM, and the right shoulder joint compartment was exposed. The supraspinatus tendon was cut off from the right proximal humerus bone. The detached end of the tendon was placed in a silicon tube (10 mm length and 4 mm diameter) and fixed by suturing (**Figure 1**). The silicon tube-inserted tendon was relocated and sutured using 5-0 Vicryl. Postoperatively, the animals were allowed free movement without the use of any type of immobilization and had access to food and water from the first day onward. They were monitored daily for 8 weeks to check their mobility and the state of the surgical wound.

2.3 Surgical procedure for dermal graft insertion

For dermal graft implantation, the supraspinatus repair by single-row repair technique was performed in this study. After 4 weeks, the silicon tube was removed from the cut supraspinatus tendon, and two holes (1 mm diameter) were made on the proximal humerus bone for the graft insertion (**Figure 2**). The rabbits were assigned to one of the following treatment groups: normal (n = 6); no grafting (n = 12), the cut supraspinatus tendon was inserted into the hole without graft and fixed by suturing; dermal graft insertion (n = 12), a sheet-shaped dermal graft (1 cm × 2 cm) was inserted and fixed by suturing between the tendon and bone; and rhBMP-2-coated dermal graft insertion (n = 12) with the same method. The animals were allowed to recover and monitored as described for chronic defect surgery, and they were sacrificed at 4 or 8 weeks.

2.4 Micro-computed tomography (CT) analysis of bone mineral density (BMD) and mineralized tissue generation

A micro-CT system was used for quantitative analysis of BMD and mineralized tissue generation between the connected tendon and bone lesion. Rabbit shoulder

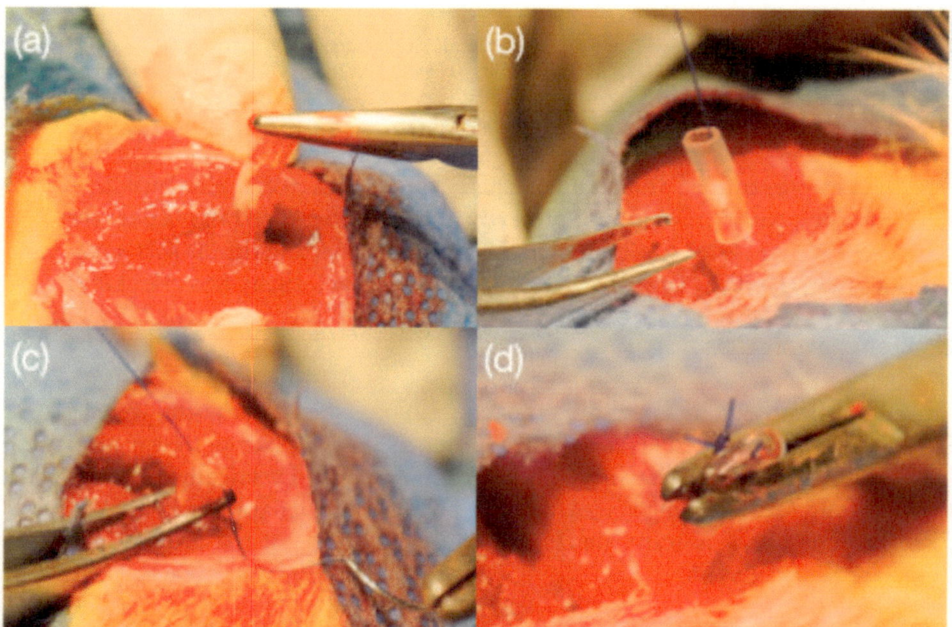

Figure 1.
The surgical procedure of chronic disease model. (a) After detaching the tendon from the proximal humeral bone; (b) the end of a tendon is inserted into the silicon tube; (c, d) they are sutured and fixed for 4 weeks.

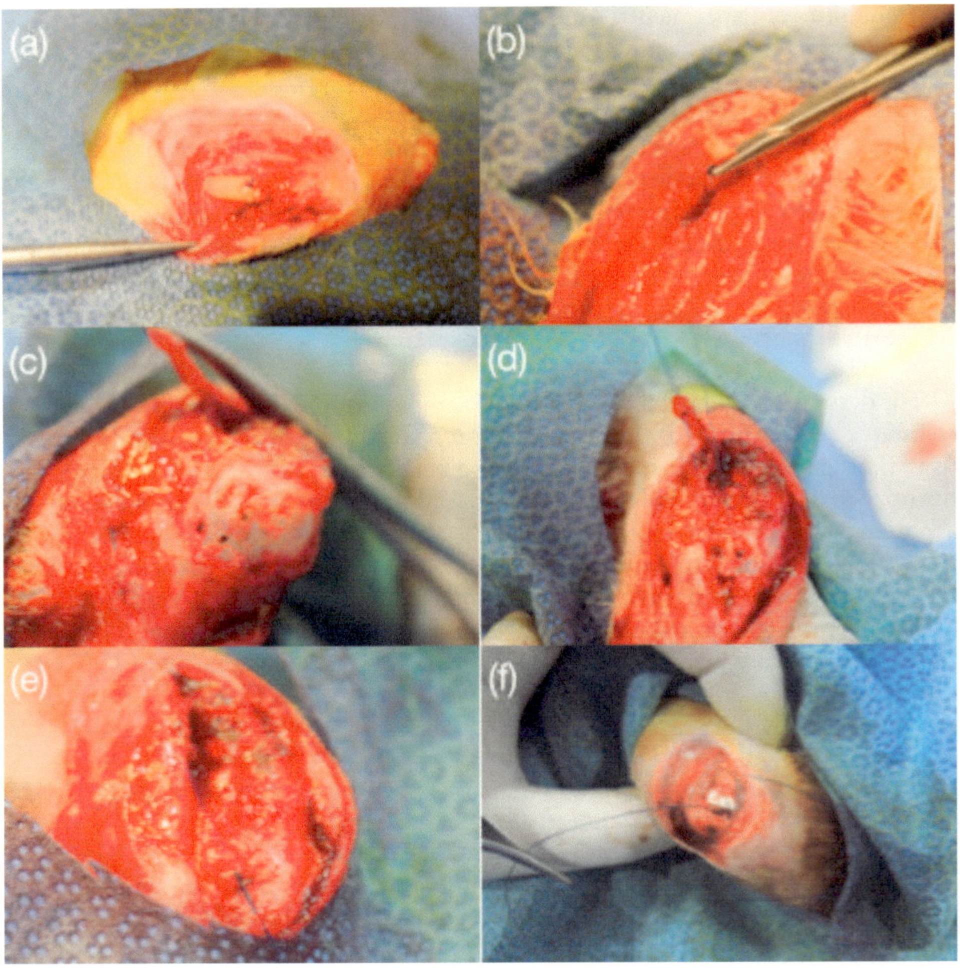

Figure 2.
The surgical procedure of patch insertion. (a) Finding the silicon tube-inserted tendon lesion after 4 weeks; (b) the removed tube from tendon; (c) making a hole for fixing with tendon; (d) the detached tendon is sutured with the bone through the holes; (e) the tendon is fixed with the bone; (f) each conditional dermal patch is inserted and fixed with suturing between the tendon and bone lesion.

specimens were scanned and reconstructed using 3D reconstruction bundle software. To quantify the amount of newly formed mineralized tissue over time, the region of interest (ROI) was chosen and reconstructed using the 3D software. After thresholding, the BMD (mm^3) of the ROI of mineralized tissue was calculated.

2.5 Biomechanical testing

Rabbit shoulder joints including the tendon-connected bone lesion were collected. To analyze the tensile mechanical properties, tensile strength was measured using a universal testing machine. The specimen was fixed vertically on a 5000 N load cell, and tensile strength was measured by pulling the sample at a load-displacement rate of 10 mm/min (**Figure 3**). The failure load and ultimate strength (N) were recorded.

2.6 Histological and histomorphometric analysis

Rabbit shoulder joints were collected and fixed in a neutralized formalin solution for 2 days and decalcified using 10% formic acid. Then, the specimens were

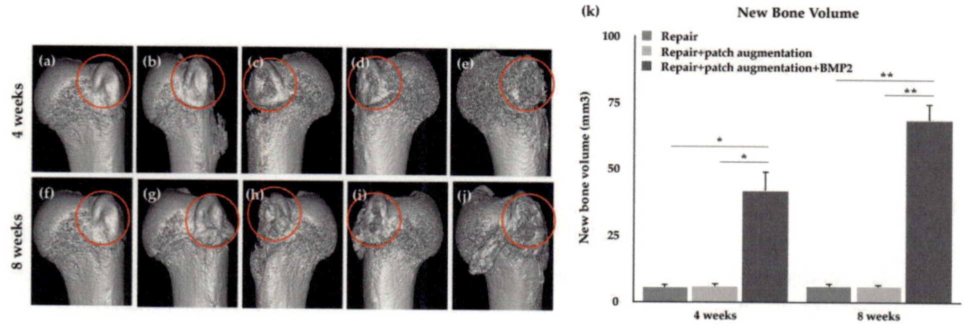

Figure 3.
3D-CT images and new bone formation of each experimental group 4 and 8 weeks after repair. (a) Intact group (PO 4 weeks); (b) defect group (PO 4 weeks); (c) repair group (PO 4 weeks); (d) dermal graft augmentation group (PO 4 weeks); (e) rhBMP-2-coated dermal graft augmentation group (PO 4 weeks); (f) intact group (PO 8 weeks); (g) defect group (PO 8 weeks); (h) repair group (PO 8 weeks); (i) dermal graft augmentation group (PO 8 weeks); (j) rhBMP-2-coated dermal graft augmentation group (PO 8 weeks); (k) quantitative analysis of the new bone formation after rotator cuff surgery after 4 and 8 weeks.

Characteristics	Points (maximum total of 9 points)
Fibrocartilage formation	
Abundant	3
Moderate	2
Slight	1
None	0
New bone formation	
Abundant	3
Moderate	2
Slight	1
None	0
Tendon allograft bonding to adjacent tissue	
75–100%	3
50–75%	2
25–50%	1
0–25%	0

Table 1.
Histomorphometric scoring system used to assess healing of the tendon-patch grafts with the bone.

dehydrated in ethanol and embedded in paraffin. The 4-um-thick sections were stained with H&E and visualized using an optical microscope. Healing of the tendon-bone connected lesion was graded histomorphologically by two observers blinded to treatment group. Histomorphometric analysis was performed to assess healing of the bone-tendon interface on the basis of three histomorphological criteria: fibrocartilage formation, new bone formation, and tendon graft bonding to adjacent tissue, which were scored on a scale of 0–3 (maximum total score of 9) in **Table 1**.

2.7 Statistical analysis

The data are presented as the average from at least triplicate samples. The experiments were repeated three times to ensure the reproducibility of the methods used. All statistical analyses were performed using SPSS (v.15.0). ANOVA was used to find overall differences among means. The post hoc Sheffé test was used to analyze differences between groups, with significance levels set at $^{*}p < 0.05$ and $^{**} < p < 0.01$.

3. Results

In the postoperative 3D-CT images, the rhBMP-2-coated dermal graft inserted group showed significantly higher new bone formation than the other groups (**Figure 3**). The new bone volume in the rhBMP-2 group at 4- and 8-week post-repair surgery was 41.8 and 68.5 mm^3, respectively, which was ~11.5 times higher than in the control and dermal graft only inserted groups. The latter groups displayed less than 6 mm^3 and, thus, did not show actual new bone formation.

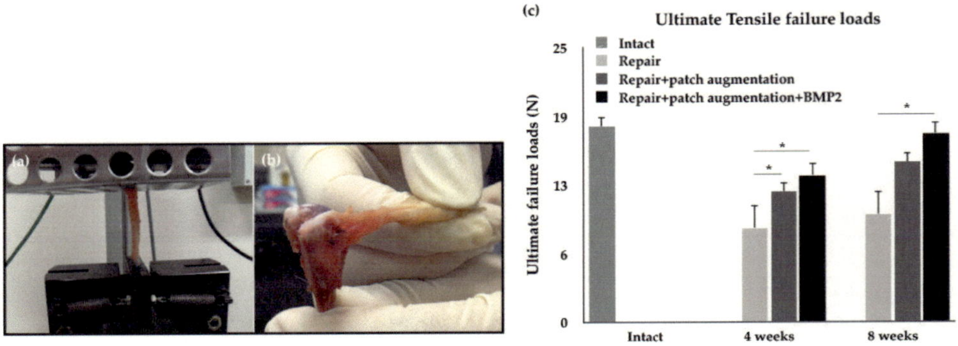

Figure 4.
The rabbit humeral bone is fixed with upper grip and the end of the detached tendon is fixed with lower grip. (a) Installed tissue specimen between the load cells for tensile test; (b) collected rabbit shoulder joint for tensile test; (c) ultimate tensile failure loads: A load-displacement rate was 10 mm/min and ultimate failure loads were measured from each group PO 4 and 8 weeks.

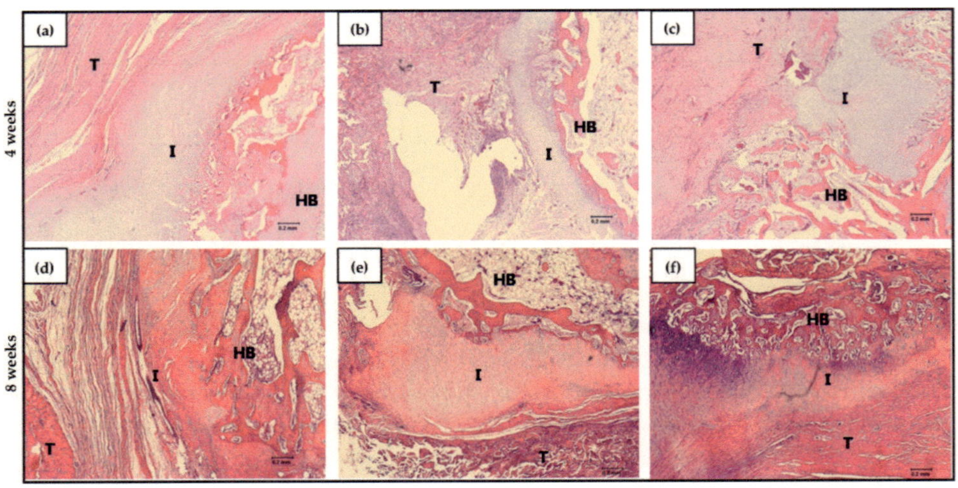

Figure 5.
The conditionally repaired groups in the rabbit tendon injury model. The repair group in which supraspinatus tendon and humeral bone were sutured at 4 weeks (a) and 8 weeks (d) post-surgery. The group that had received only a dermal patch, which covered the tendon-and-bone-suturing lesion, at 4 weeks (b) and 8 weeks (e) post-surgery. The group that had received the rhBMP-2-coated dermal patch, at 4 weeks (c) and 8 weeks (f) after repair. Images are shown at original magnification ×100. H&E, hematoxylin and eosin; T, tendon; HB, host bone; I, interface.

The rhBMP-2-coated dermal graft effectively induced new bone formation in the repaired lesion of the tendon and bone.

The ultimate tensile strength of the repaired tendon-bone interface was significantly higher in the rhBMP-2-coated dermal graft group and dermal graft only group than in the control group (suture only) at 4-week post repair surgery. However, there was a significant difference only in the rhBMP-2-coated dermal graft inserted group at 8-week post repair surgery (**Figure 4**). Moreover, the ultimate failure loads of specimens from the rhBMP-2-coated dermal graft group were similar to those of intact tissues at 8-week post-repair surgery. These results indicated not only that dermal patch provided good mechanical support but also that the rhBMP-2 increased the fusion rates of repaired lesion between the humeral bone and supraspinatus tendon.

Histological analysis indicated that newly differentiated fibrochondrocytes were detected near the host bone in the rhBMP-2-coated dermal graft group at 8-week post-repair surgery (**Figure 5F**). The dermal graft only group showed lower cell penetration into the interface area, although the dermal graft was connected between the bone and tendon (**Figure 5B, E**). However, high cell penetration was observed near the host bone surface, and the interface between the bone and tendon was well connected without any space in the rhBMP-2-coated dermal graft group (**Figure 5C, F**).

Quantitative histomorphometric analysis showed that the dermal graft coated with rhBMP-2 group had the highest analytic score (5.9) such as more fibrocartilage formation, new bone formation, and stronger attachment of the tendon to adjacent bone than the other groups (control group, 3.4; dermal graft only group, 3.9) after 8 weeks (**Figure 6**). As a result, the histological and histomorphometric analyses suggested that the rhBMP-2-coated dermal graft stimulated cell recruitment as well

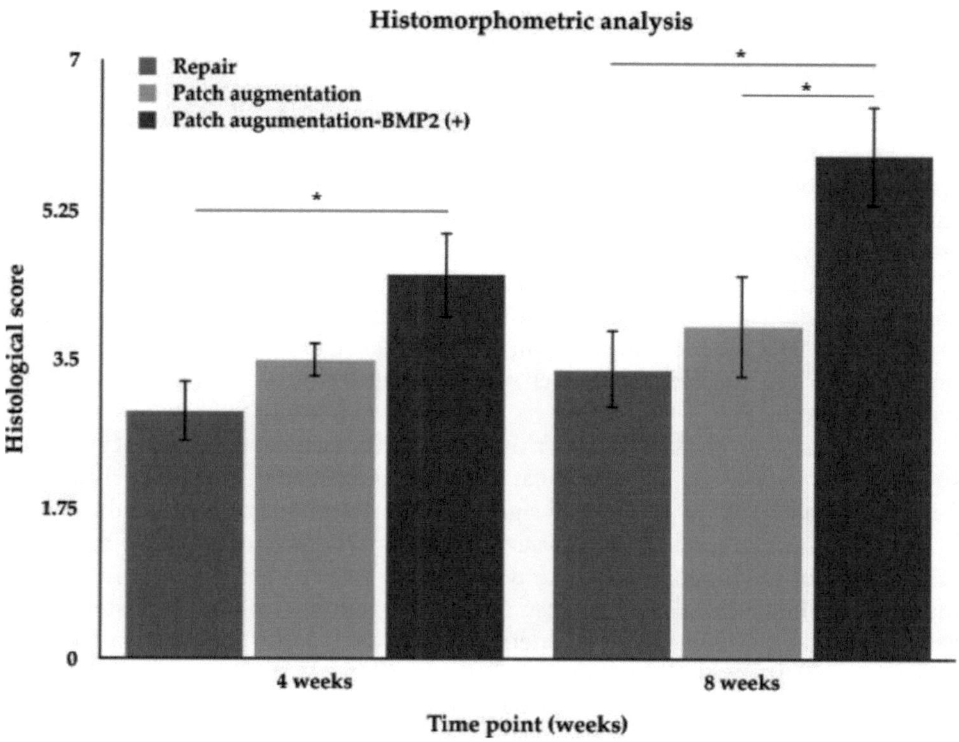

Figure 6.
Histomorphological scores of each group PO 4 and 8 weeks.

as induced fibrochondrogenic differentiation of the undifferentiated mesenchymal cells in the junction between the bone and tendon.

4. Discussion

There have been a number of preclinical and clinical trials for the enhanced tendon-bone healing in the shoulder repair surgery [12, 21, 30]. The augmentation repair of rotator cuff by incorporating growth factors or stem cells into the scaffold is intended to enhance the repair process, improve the mechanical properties, and reduce the possibility of re-rupture [10, 24, 25].

Decellularized human dermis has better mechanical properties than any other ECM-based scaffolds in the previous studies [31, 32]. ECM-based grafts are attractive model for tissue regeneration. However, they may support only short-term reinforcement and retard tissue regeneration [33]. For the enhanced healing process, the grafting material requires good mechanical properties as well as specific differentiation effects. rhBMP-2 has shown biological effectiveness for rotator cuff repair and for tendon-bone interface healing in extra-articular bone tunnel in the previous studies [28, 34]. However, there have been no preclinical and clinical studies on the rhBMP-2-coated acellular dermal graft for rotator cuff repair.

Many limitations of safety and effectiveness still remain for the clinical use of rhBMP-2. The considerations of biomechanical property for better morbidity of bone-tendon, the concentration of rhBMP-2 for the safety in the implanted area, and coating method for sustained release from the scaffold are necessary for growth factor application in the biological augmentation research.

Although the review articles suggested that dermal graft has good biomechanical properties and rhBMP-2 is effective for biological augmentation in bone-tendon, there was no original article that showed the composite of rhBMP-2 and an acellular dermal graft [35, 36]. For this initial study on rhBMP-2-coated dermal graft for rotator cuff repair, we generated a rabbit rotator cuff injury model by using a silicon tube, which covers the supraspinatus tendon that was detached from the humeral bone for 4 weeks. In our previous study, we showed the usability of rhBMP-2 application for tendon-tibia bone interface healing, using viscous collagen gel as an injectable carrier. rhBMP-2 needs to be embedded in a scaffold for an effective delivery to the targeted lesion. Thus, various types of carriers need to be tested for rhBMP-2 delivery.

Acellular dermal graft exhibits higher suture pullout strength than other ECM-based scaffolds. The rotator cuff area is subject to excessive and multidirectional loads by tendon-bone interface motion; therefore, healing following rotator cuff repair largely depends on robust mechanical properties of the junction between the bone and tendon.

On the basis of prior knowledge, rhBMP-2-coated acellular dermal graft was sutured between the rabbit supraspinatus tendon and humeral bone. The rhBMP-2-coated dermal graft group showed significant new bone formation as indicated by 3D-CT data recorded at 4 and 8 weeks after surgical treatment. The ultimate failure loads of the rhBMP-2/acellular dermal graft inserted tendon-bone interfaces increased to the mechanical properties of intact tissues at 8 weeks. The histological images demonstrated rich cell penetration into the tendon-bone interface by rhBMP-2, and the recruited cells induced fibrochondrogenic differentiation. The histomorphometric scores of the dermal graft coated with rhBMP-2 group were significantly higher than the other groups.

In conclusion, the grafting of rhBMP-2-coated human dermal graft enhanced new bone formation and the bone-tendon fusion rate in a rabbit chronic rotator cuff injury model. This study demonstrated that the combination of an acellular dermal graft and rhBMP-2 can be applied to that rotator cuff injured patients.

Author details

Kwang-Il Lee[1*], Ju-Woong Jang[2] and Kwang-Won Lee[3]

1 The Scripps Research Institute, La Jolla, CA, United States

2 The Institute of Biomaterials and Medical Engineering, Cellumed Co., Ltd., Seoul, Republic of Korea

3 Department of Orthopedic Surgery, Eulji University Hospital, Daejeon, Republic of Korea

*Address all correspondence to: jasonklee77@gmail.com

References

[1] Wikipedia. Rotator Cuff [Internet]. 2018. Available from: https://en.wikipedia.org/wiki/Rotator_cuff [Accessed: 23-10-2018]

[2] OrthoInfo. Rotator Cuff Tears [Internet]. 2018. Available from: https://orthoinfo.aaos.org/en/diseases-conditions/rotator-cuff-tears/ [Accessed: 23-10-2018]

[3] Oliva F, Osti L, Padulo J, Maffulli N. Epidemiology of the rotator cuff tears: A new incidence related to thyroid disease. Muscle, Ligaments and Tendons Journal. 2014;**4**(3):309-314

[4] UW Medicine Orthopaedics and Sports Medicine. Arthroscopic Shoulder Surgery for the Treatment of Rotator Cuff Tears [Internet]. 2018. http://www.orthop.washington.edu/?q=patient-care/articles/sports/arthroscopic-shoulder-surgery-for-the-treatment-of-rotator-cuff-tears [Accessed: 23-10-2018]

[5] Arthrex. Rotator Cuff Repair [Internet]. 2018. https://www.arthrex.com/shoulder/rotator-cuff-repair [Accessed: 23-10-2018]

[6] Medscape. AAOS 2012 Annual Meeting [Internet]. 2012. https://www.medscape.com/viewarticle/758280 [Accessed: 23-10-2018]

[7] Johns Hopkins Medicine. Orthopaedic Surgery [Internet]. 2018. https://www.hopkinsmedicine.org/orthopaedic-surgery/specialty-areas/shoulder/treatments-procedures/failed-rotator-cuff-repairs.html [Accessed: 23-10-2018]

[8] Lädermann A, Burkhart SS, Hoffmeyer P, Neyton L, Collin P, Yates E, et al. Classification of full-thickness rotator cuff lesions: A review. EFORT Open Reviews. 2017;**1**(12):420-430

[9] Dejardin LM, Arnoczky SP, Ewers BJ, Haut RC, Clarke RB. Tissue-engineered rotator cuff tendon using porcine small intestine submucosa. Histologic and mechanical evaluation in dogs. The American Journal of Sports Medicine. 2001;**29**(2):175-184

[10] Prabhath A, Vernekar VN, Sanchez E, Laurencin CT. Growth factor delivery strategies for rotator cuff repair and regeneration. International Journal of Pharmaceutics. 2018;**544**(2):358-371

[11] Chainani A, Little D. Current status of tissue-engineered scaffolds for rotator cuff repair. Techniques in Orthopaedics. 2016;**31**(2):91-97

[12] Thangarajah T, Pendegrass CJ, Shahbazi S, Lambert S, Alexander S, Blunn GW. Augmentation of rotator cuff repair with soft tissue scaffolds. Orthopaedic Journal of Sports Medicine. 2015;**3**(6):2325967115587495. DOI: 10.1177/2325967115587495

[13] Adams JE, Zobitz ME, Reach JS Jr, An KN, Steinmann SP. Rotator cuff repair using an acellular dermal matrix graft: An in vivo study in a canine model. Arthroscopy. 2006;**22**(7):700-709

[14] Ide J, Kikukawa K, Hirose J, Iyama K, Sakamoto H, Mizuta H. Reconstruction of large rotator-cuff tears with acellular dermal matrix grafts in rats. Journal of Shoulder and Elbow Surgery. 2009;**18**(2):288-295. DOI: 10.1016/j.jse.2008.09.004

[15] Thangarajah T, Henshaw F, Sanghani-Kerai A, Lambert SM, Blunn GW, Pendegrass CJ. The effectiveness of demineralized cortical bone matrix in a chronic rotator cuff tear model. Journal of Shoulder and Elbow Surgery. 2017;**26**(4):619-626. DOI: 10.1016/j.jse.2017.01.003

[16] Derwin KA, Badylak SF, Steinmann SP, Iannotti JP. Extracellular matrix scaffold devices for rotator cuff repair. Journal of Shoulder and Elbow Surgery. 2010;**19**(3):467-476. DOI: 10.1016/j.jse.2009.10.020

[17] Sessions WC, Lawrence RL, Steubs JT, Ludewig PM, Braman JP. Thickness of the rotator cuff tendons at the articular margin: An anatomic cadaveric study. The Iowa Orthopaedic Journal. 2017;**37**:85-89

[18] Bello YM, Falabella AF, Eaglstein WH. Tissue-engineered skin. Current status in wound healing. American Journal of Clinical Dermatology. 2001;**2**(5):305-313

[19] Schmidt J, Redwan B, Koesek V, Heitplatz B, Bedetti B, Aebert H, et al. Thoracic wall reconstruction with acellular porcine dermal collagen matrix. The Thoracic and Cardiovascular Surgeon. 2016;**64**(3):245-251. DOI: 10.1055/s-0034-1383512

[20] Eppley BL, Summerlin DJ, Sadove AM. A potential biomaterial composite for dermal and subcutaneous augmentation. Annals of Plastic Surgery. 1994;**32**(5):463-468

[21] Edwards SL, Lynch TS, Saltzman MD, Terry MA, Nuber GW. Biological and pharmacologic augmentation of rotator cuff repairs. The Journal of the American Academy of Orthopaedic Surgeons. 2011;**19**(10):583-589

[22] Bedi A, Maak T, Walsh C, Rodeo SA, Grande D, Dines DM, et al. Cytokines in rotator cuff degeneration and repair. Journal of Shoulder and Elbow Surgery. 2012;**21**(2):218-227

[23] Wd G, Xu PC, Wu P, Hu K, Sun Y, Cheng B, et al. Advances in the treatment of rotator cuff lesions by cytokines. Frontiers in Bioscience (Landmark Edition). 2017;**22**:516-529

[24] Cheung EV, Silverio L, Yao J. Delivered growth factor therapy to improve healing after rotator cuff repair. Stem Cells Cloning. 2010;**3**:135-144. DOI: 10.2147/SCCAA.S7359

[25] Oliva F, Via AG, Maffulli N. Role of growth factors in rotator cuff healing. Sports Medicine and Arthroscopy Review. 2011;**19**(3):218-226. DOI: 10.1097/JSA.0b013e3182250c78

[26] Rodeo SA, Suzuki K, Deng XH, Wozney J, Warren RF. Use of recombinant human bone morphogenetic protein-2 to enhance tendon healing in a bone tunnel. American Journal of Sports Medicine. 1999;**27**(4):476-488

[27] Hashimoto Y, Yoshida G, Toyoda H, Takaoka K. Generation of tendon-to-bone interface "enthesis" with use of recombinant BMP-2 in a rabbit model. Journal of Orthopaedic Research. 2007;**25**(11):1415-1424

[28] Lee KW, Lee JS, Jang JW, Shim YB, Lee KI. Tendon-bone interface healing using an injectable rhBMP-2 containing collagen gel in a rabbit extra-articular bone tunnel model. Journal of Tissue Engineering and Regenerative Medicine. 2017;**11**(5):1435-1441. DOI: 10.1002/term.2041

[29] Lee KI, Lee JS, Kim JG, Kang KT, Jang JW, Shim YB, et al. Mechanical properties of decellularized tendon cultured by cyclic straining bioreactor. Journal of Biomedical Materials Research. Part A. 2013;**101**(11):3152-3158. DOI: 10.1002/jbm.a.34624

[30] Atesok K, Fu FH, Wolf MR, Ochi M, jazrawi LM, Doral NM, et al. Augmentation of tendon-to-bone healing. The Journal of Bone and Joint Surgery. American Volume. 2014;**96**(6):513-521. DOI: 10.2106/JBJS.M.00009

[31] Omae H, Steinmann SP, Zhao C, Zobitz ME, Wongtriratanachai P, Sperling JW, et al. Biomechanical effect of rotator cuff augmentation with an acellular dermal matrix graft: A cadaver study. Clinical Biomechanics (Bristol, Avon). 2012;**27**(8):789-792. DOI: 10.1016/J.clinbiomech.2012.05.001

[32] Nie X, Wang DA. Decellularized orthopaedic tissue-engineered grafts: Biomaterial scaffolds synthesized by therapeutic cells. Biomaterials Science. 2018;**24**(6):2798-2811. DOI: 10.1039/c8bm00772a

[33] Sackett SD, Tremmel DM, Ma F, Feeney AK, Maguire RM, Brown ME, et al. Extracellular matrix scaffold and hydrogel derived from decellularized and delipidized human pancreas. Scientific Reports. 2018;**8**(1):10452. DOI: 10.1038/s41598-018-28857-1

[34] Lee KW, Lee JS, Kim YS, Shim YB, Jang JW, Lee KI. Effective healing of chronic rotator cuff injury using recombinant bone morphogenetic protein-2 coated dermal patch in vivo. Journal of Biomedical Materials Research. Part B, Applied Biomaterials. 2017;**105**(7):1840-1846. DOI: 10.1002/jbm.b.33716

[35] Acevedo DC, Shore B, Mirzayan R. Orthopedic applications of acellular human dermal allograft for shoulder and elbow surgery. The Orthopedic Clinics of North America. 2015;**46**(3):377-388. DOI: 10.1016/j.ocl.2015.02.006

[36] Kim JG, Kim HJ, Kim SE, Bae JH, Ko YJ, Park JH. Enhancement of tendon-bone healing with the use of bone morphogenetic protein-2 inserted into the suture anchor hole in a rabbit patellar tendon model. Cytotherapy. 2014;**16**(6):857-867. DOI: 10.1016/j.jcyt.2013.12.012

Chapter 6

Structural Design, Fabrication and Evaluation of Resorbable Fiber-Based Tissue Engineering Scaffolds

Martin W. King, Jiyang Chen, Monica Deshpande, Ting He, Harshini Ramakrishna, Yu Xie, Fan Zhang and Fan Zhao

Abstract

The use of tissue engineering to regenerate viable tissue relies on selecting the appropriate cell line, developing a resorbable scaffold and optimizing the culture conditions including the use of biomolecular cues and sometimes mechanical stimulation. This review of the literature focuses on the required scaffold properties, including the polymer material, the structural design, the total porosity, pore size distribution, mechanical performance, physical integrity in multiphase structures as well as surface morphology, rate of resorption and biocompatibility. The chapter will explain the unique advantages of using textile technologies for tissue engineering scaffold fabrication, and will delineate the differences in design, fabrication and performance of woven, warp and weft knitted, braided, nonwoven and electrospun scaffolds. In addition, it will explain how different types of tissues can be regenerated by each textile technology for a particular clinical application. The use of different synthetic and natural resorbable polymer fibers will be discussed, as well as the need for specialized finishing techniques such as heat setting, cross linking, coating and impregnation, depending on the tissue engineering application.

Keywords: resorbable polymer, tissue engineering scaffold, biocompatibility, cell culture, porosity, braiding, knitting, weaving, nonwoven web

1. Introduction

The field of tissue engineering and regenerative medicine was conceptualized about 35 years ago by Robert S. Langer, Institute Professor at Massachusetts Institute of Technology, together with Joseph P. Vacanti, MD a pediatric and transplantation surgeon-scientist at Massachusetts General Hospital and Harvard Medical School [1]. They recognized that in all fields of reconstructive surgery less than 30% of patients needing an organ transplant were able to obtain one [2], and that an alternative approach of engineering viable tissues and organs using cell culture techniques was needed to address this limited supply [3]. As a result, "an interdisciplinary field of research that applies the principles of engineering and the life sciences towards the development of biological substitutes that restore, maintain or improve tissue function" has been created with the ultimate goal of repairing injured and diseased organs [1].

Originally, the concept of tissue engineering required a triad of cells, scaffolds and signaling molecules. The cells, preferably derived autologously from the patient, can be stem cells, progenitor cells or mature cells. After expansion, they are seeded on a porous and resorbable scaffold and cultured in the presence of signaling molecules such as growth factors, specific metabolites, morphogens and adhesins [4]. Today, it is known that additional conditions are required so as to obtain the desired tissue. Depending on the type of tissue and the application, it may be necessary to add mechanical [5] and/or bioelectrical stimulation [6, 7], or increase the oxygen levels during cell culture in the bioreactor [8].

The focus of this chapter is to describe the different types of scaffolds used for tissue engineering with a particular emphasis on fiber-based scaffolds that are mostly fabricated on textile processing equipment. The desirable properties of the various types of scaffolds will be enumerated and the specific advantages of using fiber-based structures will be explained.

2. Specifications for tissue engineering scaffolds

Figure 1 represents the desired structural, mechanical and chemical properties that every tissue engineering scaffold should possess. The structure needs to have open pores that facilitate the passage of nutrients and waste products to and from the cells in the interior of the scaffold. The average pore size and pore size distribution are critical so as to ensure that cells, which may measure 10–50 μm in size, can infiltrate into the interior of the scaffold. This is particularly crucial for endothelial cells that are responsible for the formation of vasa vasorum and internal vascularization of the scaffold tissue.

The surface properties of the scaffold should be suitable from both chemical and topographical points of view so as to enable cell attachment and proliferation. To this end surfaces are often coated or modified with extra cellular matrix proteins, such as collagen or fibronectin, to promote cellular interaction. It is assumed that all surfaces are biocompatible so as elicit cell interaction and avoid a cytotoxic response.

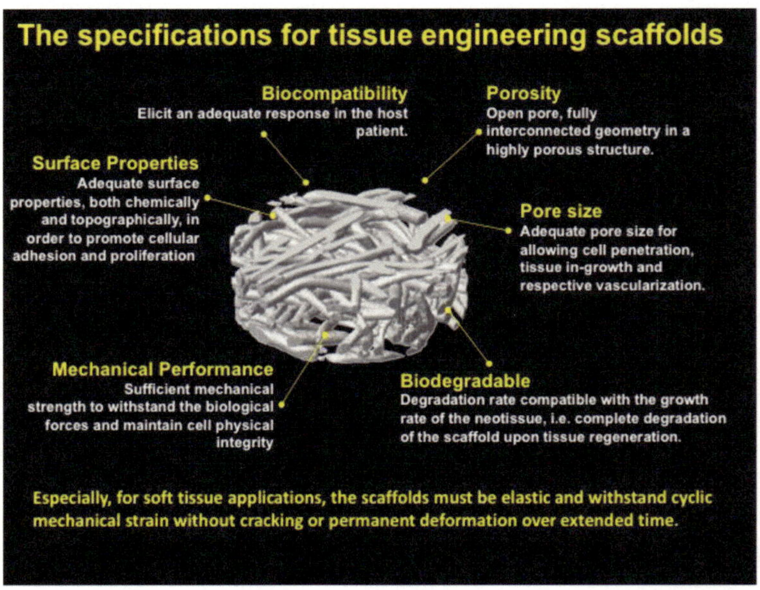

Figure 1.
Required properties of tissue engineering scaffolds.

3. Types of nonfiber-based tissue engineering scaffolds

The scaffold structures fabricated by a number of conventional methods, such as solvent casting, particulate leaching, gas foaming, phase separation and freeze drying are illustrated in **Figure 2**. The porous structures are inconsistent and there are limited processing options to alter the average pore size and the pore size distribution. Hence the mechanical performance in any direction is limited, especially for scaffolds that need to support mechanical loading during cell culture. In addition, a number of the polymer materials require the use of toxic organic solvents that increases the risk of cytotoxic effects during cell culture.

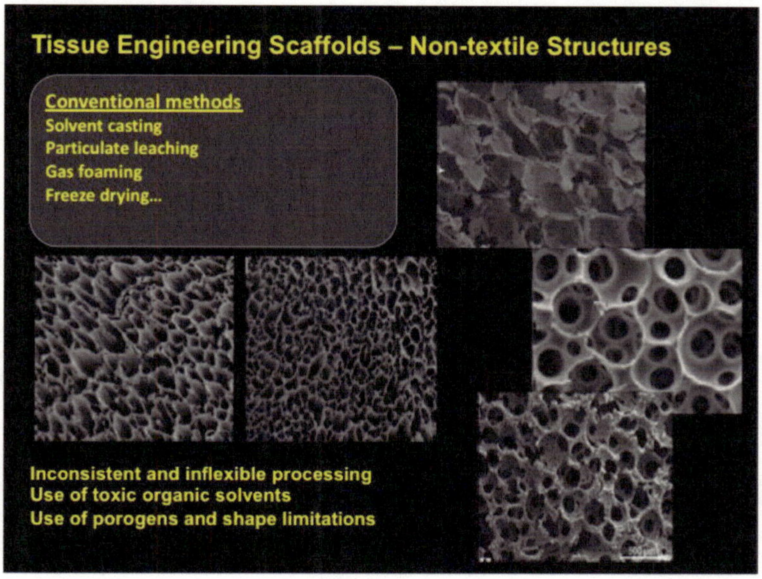

Figure 2.
Porous structures of non-textile tissue engineered scaffolds.

4. Types of fiber-based tissue engineering scaffolds

As seen in **Figure 3**, there a number of textile technologies that can and have been used to fabricate tissue engineering scaffolds. They include weaving, weft knitting, warp knitting, braiding, nonwoven production and electrospinning. Each of these technologies is described in the following sections, which discuss both the concepts and principles as well as examples of their application.

4.1 Woven scaffolds

Weaving is a conventional textile fabrication technology that is widely used in tissue engineering applications, because it enables a 3D scaffold to be fabricated that can imitate the mechanical and biological features of native human tissues. Woven structures are basically formed by interlacing two sets of yarns, namely, warp and weft yarns. With various interlacing and pattern designs, woven fabrics are categorized as plain, twill and satin weaves (**Figure 4**). Compared with knitted and braided structures, woven fabrics have better mechanical strength and structural stability [9]. The properties of woven fabrics, such as thickness, porosity and strength, can be easily adjusted and modified by woven design selection and the density of the warp and weft yarns [10].

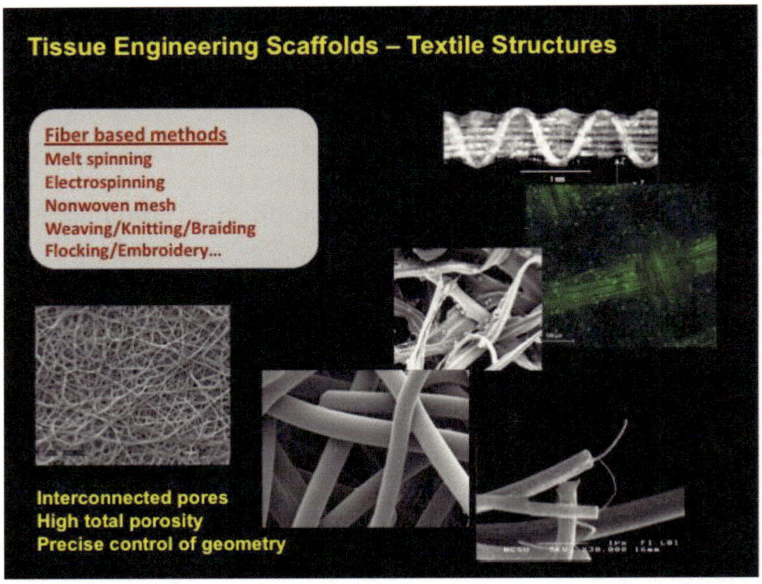

Figure 3.
Porous structures of fiber-based tissue engineered scaffolds.

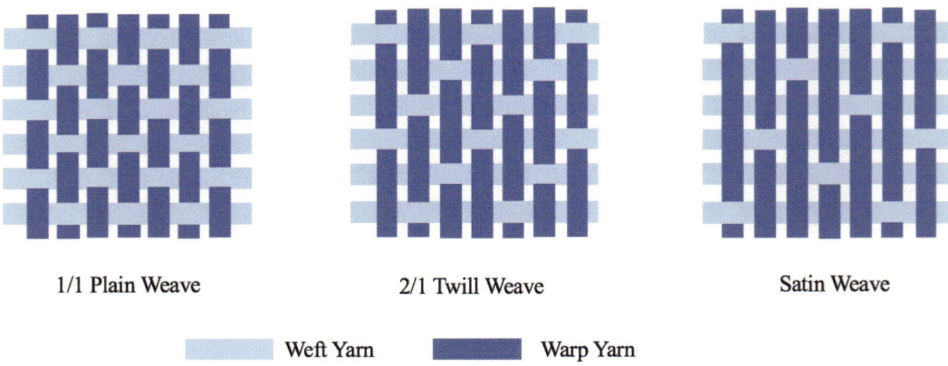

Figure 4.
Woven designs showing plain, twill and satin weaves [10].

4.1.1 Woven tendon bridging and reinforcement

A tendon is a crucial linkage between a muscle and bone, and it plays an important role in the movement of the joints, such as the rotator cuff tendon which allows shoulder movement. When a tendon ruptures or tears, it causes dysfunction of the joint. The complex multilayered avascular structure limits the rate and potential for healing [11–13]. Tissue engineered scaffolds not only need to have excellent mechanical properties such as high tensile stress and modulus, but also should have excellent biological properties to promote rapid host cell growth and tissue regeneration.

Commercial woven scaffolds, such as X-Repair (Synthasome, CA, USA) and Biofiber™ CM (Tornier, MN, USA) made from biodegradable polymers, are already used clinically for tendon repair. Derwin et al. reported using the woven poly-L-lactic acid (PLLA) X-Repair patch to provide bridging reinforcement for a shoulder tendon in a canine model. Post-operatively the augmented and repaired tendon was 23% significantly stronger, and after 12 weeks the patch reinforcement showed less tendon retraction and significantly greater stiffness (26%), and ultimate load (35%) compared to those animals that were repaired without a patch [14]. For the first time in 2014, Proctor successfully used the X-Repair woven patch with an arthroscopic approach to repair a series of patients who presented with a large-to-massive rotator cuff tear. He reported that the surgery provided substantial functional improvement for 83% of patients after 12 months and 78% of the patients after 42 months [15]. Ratcliffe et al. compared the mechanical properties of a number of commercially available synthetic and extracellular matrix (ECM) scaffolds, and found that only the X-Repair was able to provide a similar stress-strain curve to a human or canine infraspinatus rotator cuff tendon with a short toe region and high strength (**Figure 5**) [16].

More recently, a woven structure has been used together with non-woven layers to assemble a laminated multilayered scaffold (**Figure 6A**) to closely match the mechanical properties of a human rotator cuff tendon and provide nanofiber *in vitro* and *in vivo* bioactivity [17]. Wu et al. combined the nanofibers with a woven structure by weaving the nanofiber yarn directly into the scaffold. With dynamic conditioning, the scaffold can promote significant collagen secretion and tenogenic differentiation of the tri-culture of derived mesenchymal stem cells, human tenocytes, and human umbilical vein endothelial cells [18].

Islam et al. manually wove plied and crosslinked electrochemically aligned collagen threads (ELACs) using pins into a scaffold for rotator cuff tendon repair

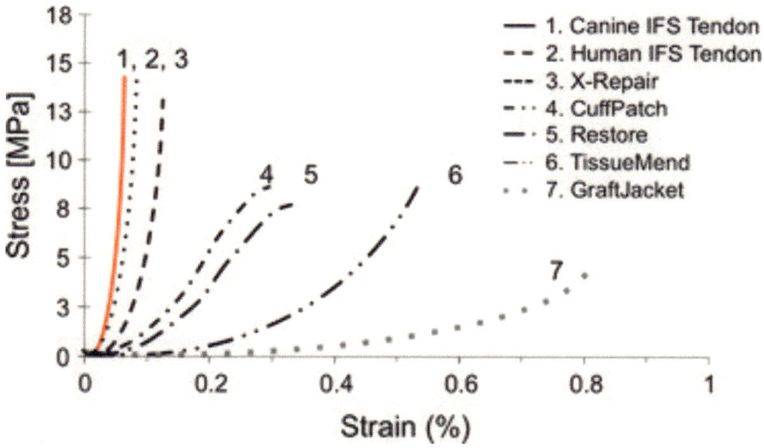

Figure 5.
Stress–strain curves for the human and canine infraspinatus rotator cuff tendon, and for the products used in the repair of rotator cuff injuries [16].

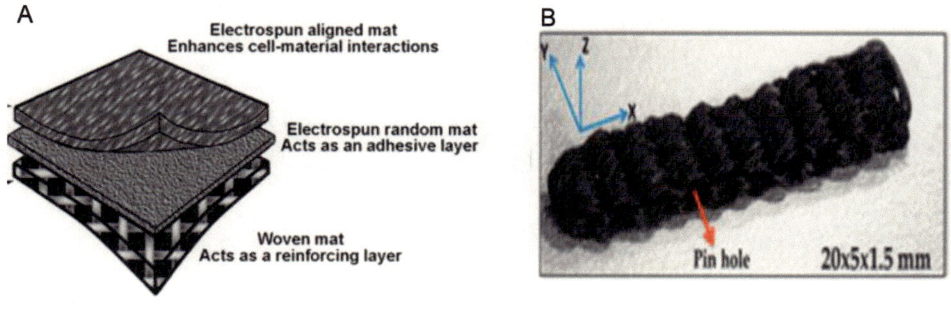

Figure 6.
Recent woven scaffold innovations for tendon repair applications. (A) Composite multilayer woven and electrospun scaffold [17]. (B) Pin woven scaffold made from novel electrochemically aligned collagen threads [19].

(**Figure 6B**) [19]. The scaffold has approximately 60% of the functional strength of a comparable sized native rabbit infraspinatus tendon, a stiffness close to that of a native tendon and the ability to initiate tenogenic differentiation of human mesenchymal stem cells [20].

4.1.2 Bone and cartilage regeneration

Human bone has a complex hierarchical and lamella structure of mineralized collagen fibrils which makes it difficult to replicate the complex ECM structure for better cell growth and bone regeneration [21]. By adjusting the pore size and structure of the woven fabric a potential scaffold candidate can address the required mechanical and biological features of an ideal tissue engineering bone scaffold [22]. Recently, a three dimensional (3D) woven structure has shown potential for bone regeneration.

Polylactic acid (PLA) and silk fibroin were combined and electrospun into nanofibers and fabricated into a woven multilayer fabric with subsequent mineralization using simulated body fluid. This approach significantly improved the scaffold's compression resistance and enhanced cell proliferation by promoting osteogenic differentiation of mesenchymal stem cells (MSC) [21]. This stem cell differentiation was also confirmed by Persson et al. who seeded cells onto a 2.4 mm thick 3D woven scaffold made from wet spun PLA and hydroxyapatite (HA) composite fibers (**Figure 7**) [23]. A 3D engineered woven poly(ε-caprolactone) (PCL) scaffold was created for the purpose of assisting deposition of cartilaginous and mineralized matrix from marrow-derived human bone for repairing chondral or osteochondral defects [24]. By using subchondral bone anchor in a porcine *in vivo* model, a 3D PCL woven scaffold has the potential to be used for long-term repair of chondral defects [25], and with an MSC-seeded hydrogel layer, the PCL woven scaffold could provide a microenvironment for stem cell chondrogenesis [26]. Other degradable polymers have also been investigated to fabricate 3D woven scaffolds for bone regeneration, such as poly-L-lactic acid (PLLA) [27].

4.1.3 Aortic heart valve replacement

With the problems of aortic valve insufficiency and stenosis, patients need heart valve replacement surgery to regulate blood flow between the left ventricle and the aorta [10]. The ideal scaffold should be able to simulate the native valve in terms of an active change of shape, size and stiffness of the cusp, annulus, sinus, and sinotubular junction during the cardiac cycle [28]. Textiles have been used to fabricate

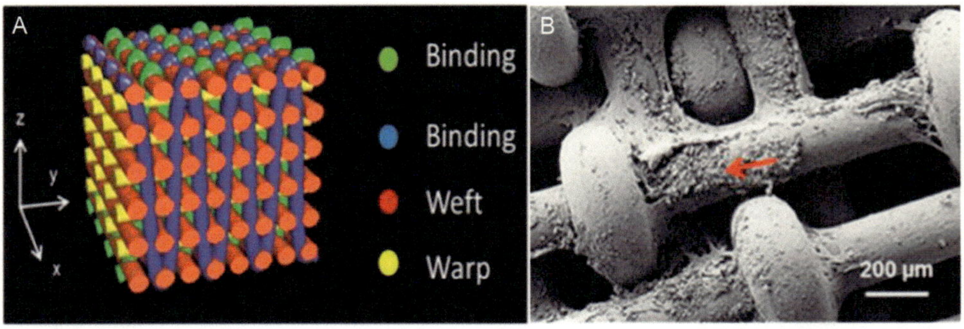

Figure 7.
(A) Schematic view of a 3D orthogonal woven scaffold, the PLA/HA 3D woven scaffold has five warp layers (x-direction) and six weft layers (y-direction), bound together by a warp set through-the-thickness (z-direction) and (B) cell mineralization (arrow) occurred on PLA/HA 3D woven scaffold after 35-days of culture in osteogenic induction medium [23].

heart valves because of their unique structural and mechanical characteristics that enable the production of unique anisotropic properties [10].

Wu et al. proposed a novel engineered valve design of combining a woven fabric with a hydrogel to mimic the heterogeneous and anisotropic features of native heart valves [29]. The scaffold was composed of a polyacrylonitrile (PAN) electrospun weft yarn and a multifilament PAN warp yarn with a methacrylated hyaluronic acid (Me-HA) or methacrylate gelatin (Me-Gel) layer for encapsulation of human aortic valve interstitial cells. The composite scaffold showed a similar initial toe, transient, and peak tangent regions in a stress-strain curve under load, similar to that of native aortic valve leaflets. High cell viability and layered cell penetration through the fibrous network were obtained after 14-days of *in vitro* culture. The woven fabric structure of the composite was able to promote the alignment and natural proliferation of normal cells, promote ECM remodeling, increase cell proliferation on the Me-HA/Me-Gel hydrogels and reduce the extent of shrinkage. Moreover, the composite scaffolds were able to suppress the trans-differentiation of diseased cells. This concept has suggested the possibility of woven structures being applied to heart valve regeneration if combined with the appropriate selection of a degradable polymer, such as PCL [10].

4.1.4 Vascular substitution and reconstruction

When patients suffer from atherosclerosis or an aneurysm, they may need an arterial prosthesis or stent-graft to replace or bypass the occluded or dilated vessel. Woven textiles have a long history of being used to fabricate arterial protheses with non-degradable synthetic polymer yarns, such as polyester, for long-term therapy of large caliber vessels. For tissue engineered vascular grafts, woven structures are less frequently used compared to other textile structures with more flexibility, such as electrospun non-woven webs and knitted structures [30–34]. However, weaving technology is capable of producing complex branched tubular structure for vascular application, such as seamless bifurcated or trifurcated endovascular prostheses [35].

Yokota has proposed a small diameter vascular graft with inner diameter equal to 4 mm, fabricated by combining a collagen microsponge with a plain-woven tube made from sheath-core yarns, namely, polyglycolic acid (PGA) as the sheath and poly-L-lactic acid (PLLA) as the core (**Figure 8A**) [36].

After 12 months of implantation in a canine carotid animal model, the graft showed no evidence of thrombogenic activity or aneurysm formation. Instead there was evidence of excellent in situ tissue regeneration [36]. The same group then examined the reconstruction potential of the vessel wall by implanting a woven patch into the canine pulmonary artery (**Figure 8B** and **C**). Similarly, no aneurysm and thrombus formation were observed after 6 months but a monolayer of

Figure 8.
(A) The small diameter vascular graft with plain weave fabricated from bicomponent yarn with PLLA core fibers and PGA sheath fibers [36]; (B) woven scaffold design with PGA/PLLA sheath-core yarns; and (C) intraoperative view of patch (arrow) implanted in the canine pulmonary artery [37].

endothelial cells and layers of smooth muscle cells were presented [37]. The ability to promote *in situ* autologous vascular regeneration with a tissue engineered PLA woven/PGA knitted/collagen-microsponge composite scaffold was also confirmed by using a porcine descending aorta model for a 6-month implantation study [38].

4.2 Weft knitted scaffolds

Weft knitting technology dates back to the sixteenth century when the Reverend William Lee in England invented a knitting frame to produce woolen hosiery. Weft knitted fabrics can include three main types of stitches: jersey, rib and interlock structures, which can be fabricated from a single yarn [39].

Compared with weaving and warp knitting, weft knitting has superior compliance and flexibility of design. When used with advanced biodegradable materials, such as poly-L-lactic acid (PLLA), polyglycolic acid (PGA) and polyurethanes (PU), their mechanical performance is predictable whether or not they serve as the reinforcing component of a tissue engineering scaffold [40, 41]. Structural instability and fabric permeability are the two main concerns relating to the performance of weft knitted fabrics. The use of advanced coatings and immersion techniques compensate for these limitations, and the combined scaffolds provide improvements in device function [42, 43]. To date, various implantable scaffolds have been fabricated as weft knitted structures, including a cardiac support device (CSD), aortic valves, vascular prostheses and nerve guides [42].

4.2.1 Cardiac support device

In the event of a heart attack or myocardial infarction (MI), the muscle wall of the left ventricle will experience a remodeling process [44]. Today, the major challenge is how to provide the mechanical support for the left ventricle and deliver stem cells to the target area where the infarct occurred. For this application, the scaffold needs to be extensible and match the different compliances of the left ventricle in the radial and longitudinal directions [45].

Until now, Boublick et al. developed a knitted cardiac patch that has been developed and assessed by *in vitro* and *in vivo* tests for its mechanical support and its ability to reverse the remodeling process [46]. This cardiac patch was knitted from hyaluronan benzyl ester (HBE) sutures (Hyaff-11®) and seeded with rat cardiac cells mixed with fibrin (**Figure 9**). This knitted, biodegradable construct exhibited excellent

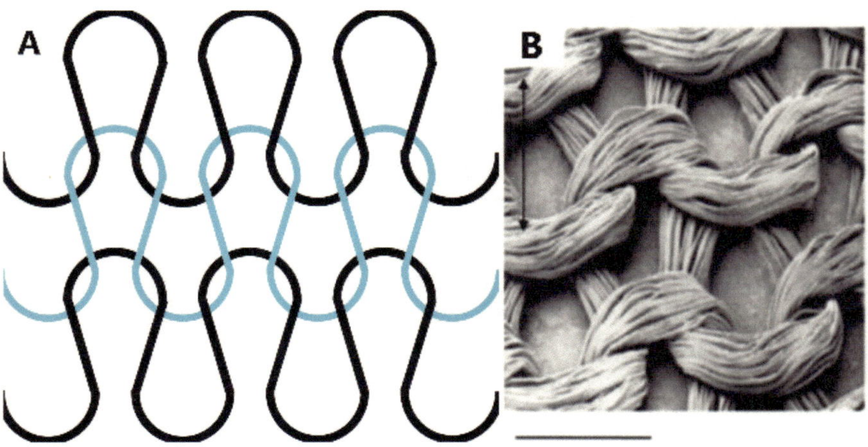

Figure 9.
(A) Three courses of weft knitted loop structure and (B) SEM image of knitted heart patch. Scale bar: 500 μm [46].

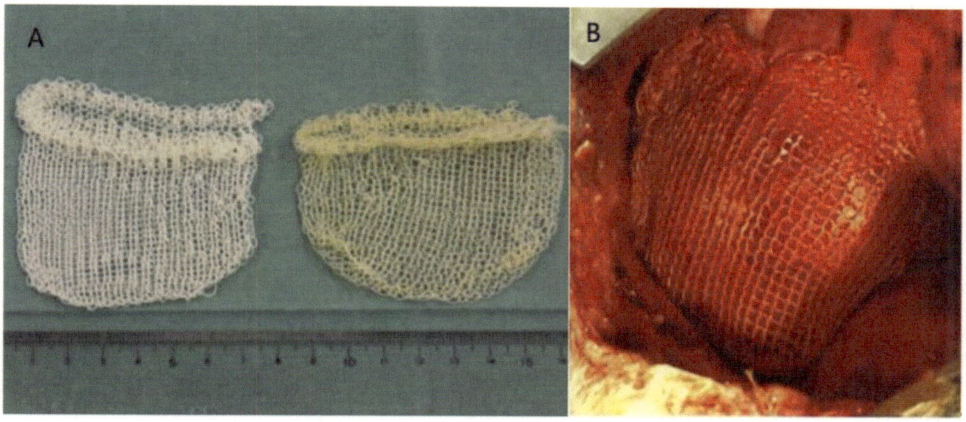

Figure 10.
Knitted cardiac support devices. (A) Non-degradable PET knitted device (left) and a degradable PGA knitted device (right) and (B) The implanted device after implantation around the ventricles of the canine heart [48].

mechanical matching with the native rat myocardium, and it was able to successfully repair the mechanical defect and malfunction of the rat heart. Chen et al. has also fabricated a PLA cardiac patch with rib stitch structure to deliver cardiosphere-derived cells (CDCs) to the ventricular wall. The *in vitro* tests indicated that (1) the mechanical properties of the rib stitch patch structure matched those of native tissues, and (2) the CDCs were able to attach to and proliferate on the patch [47].

Related to the heart patch concept, a biodegradable knitted heart cap has proven to have a beneficial effect when used in a canine heart failure model (**Figure 10**). After comparing the healing response of a biodegradable (PGA) heart cap with that of a non-degradable (PET) heart cap in a canine cardiac model, Kitahara et al. pointed out that the failure of the CorCap™ cardiac support device was that it was knitted from a non-degradable material, which was polyester (polyethylene terephthalate) (PET) yarns. The improved heart function of using biodegradable PGA in a canine model has supported this conclusion, which is consistent with the clinical data for the CorCap™ device [48, 49].

4.2.2 Aortic valve

Surgical aortic valve replacement (AVR) is conducted on patients with aortic valve stenosis [50]. The expectancy of the replacement valve to continue to function is from 10 years to a lifetime, which requires the implanted valves to have excellent durability and biocompatibility. The most challenging aspect of the surgery is the risk of post-surgical paravalvular leak, increasing the risk of the heart failure and other severe complications [51, 52]. Lieshout et al. developed a knitted scaffold for the aortic valve from polycaprolactone yarn. The fabric was knitted into a rectangular patch with three leaflets (**Figure 11**).

During valve opening and closing, the knitted leaflets performed well with complete coaptation and no collison with the aortic wall. In an *in vitro* durability fatigue test, the knitted valve experienced 10 million cycles without tearing and failing. A 3 layer coating of fibrin gel was applied to improved the biocompatibility [53]. It exhibited excellent durability and structural stability in a long-term *in vitro* fatigue study. In the future, PCL may be a good choice to regenerate heart valve leaflets in a bioreactor using native cells and extracellular matrix. This approach is likely to be more reliable than the permanent synthetic implants on the market [54].

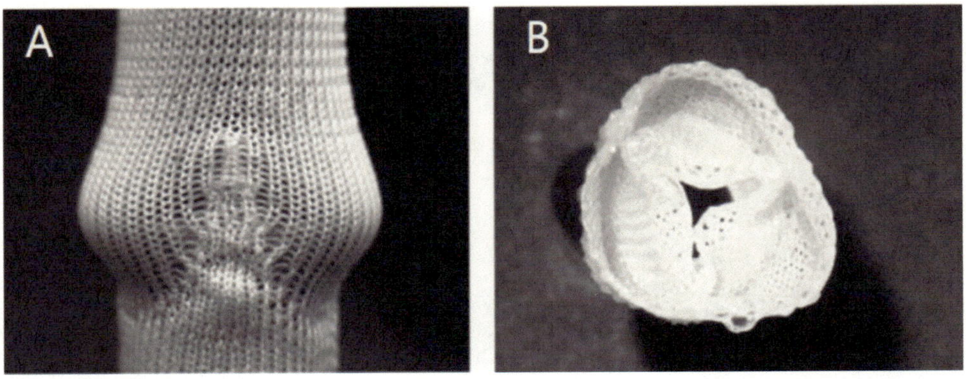

Figure 11.
The appearance of a weft knitted heart valve. (A) External lateral view and (B) top view of heart valve after application of fibrin gel layers [53].

4.2.3 Vascular prosthesis

A vascular graft is a common and useful therapy to rebuild small diameter blood vessels and treat the symptoms of atherosclerotic vascular disease [55]. However, to date only yarns spun form permanent or non-degradable polymers have been used to fabricate commercial large caliber prostheses.

With the objective of developing a resorbable small diameter vascular prosthesis, Xie et al. designed a weft knitted/electrospun PLA/PLCL tube with comparable compliance to a human saphenous vein [33]. Zhang et al. has also fabricated a small diameter vascular scaffold by weft knitting electrochemically aligned collagen (ELAC) yarns into a tubular structure on a circular weft knitting machine (**Figure 12**).

The tube serves as the backbone of the scaffold and provides sufficient mechanical support and structural integrity. Also, by utilizing collagen as the material, the scaffold has demonstrated advanced endothelial cell adhesion [56, 57]. Although a common complaint of weft knitted structures has been their poor dimensional stability and their tendency to unravel from the unsecured end, these are surmountable issues that can be controlled by knitting a backloop binding-off structure or sewing a reinforcing edge seam along the cut edge.

Figure 12.
The circular knitting machine for weft knitting tubular fabric.

4.2.4 Nerve guide

A novel approach to fabricating a nerve guide is to combine weft knitting technology with freeze drying to form a high porosity scaffold. After being injured, the distal stump of the peripheral nerve is unable to repair the gap between neurons. So to bridge this gap, a nerve guide with good compliance can be used to evaluate the neural signals and promote axonal growth [58]. Wang et al. designed and fabricated a chitosan scaffold for nerve tissue engineering. The knitted chitosan tubular fabric was fitted onto a mandrel and immersed in a chitosan solution. Acupuncture needles were then inserted into the hollow chitosan tube to create inner pores for guiding neural growth and proliferation. Following the immersion process, the scaffolds were freeze dried to form interconnected micropores in both the outer wall and inner matrix. The knitted scaffold provided sufficient compressive resistance and recovery to serve as a nerve guide scaffold. A porous microstructure improved the axonal elongation and migration of the neural cells [59]. *In vitro* evaluation of this knitted, freeze dried scaffold confirmed that it had suitable mechanical properties, and it provided good cell affinity, porosity and rate of biodegradation for neural tissue engineering [60].

4.3 Warp knitted scaffolds

Warp knitted fabrics are formed by wales which are vertical columns of yarns looped in the warp or machine direction. Warp knitted fabrics gain popularity in many medical applications due to the superior structural stability, the avoidance of yarn raveling after cutting to size, and higher suture retention strength in comparison with weft knitted fabrics.

The knitting productivity of warp knitted fabrics is usually much higher than for weft knitted fabrics. However, the yarn preparation for warp knitting is more challenging due to yarn beam preparation. There are fewer design pattern options for warp knitted fabrics than for weft knits because the warp knitted design is limited by the pattern drum on the machine, which is more difficult to create complex structures [61]. Two dimensional (2D) warp knitted fabrics have been widely adopted for biomedical applications due to their superior structural stability and durability. For example, most permanent hernia repair meshes are warp knitted so as to provide high tear resistance and bursting strength, reliable stabilization of the fascial tissue in the abdominal wall, no raveling when cut to size and limited contraction during healing [62]. Other permanent applications involve commercial arterial prostheses, aortic valve rings [53] and artificial skin [63].

On the other hand, only a few resorbable warp knitted fabrics are being developed as tissue engineering scaffolds. Secant Medical LLC (Telford, PA, USA) is developing a warp knitted fabric using degradable yarns for a tissue engineering application [64]. In order for a tissue engineering scaffold to mimic the volume and complexity of natural tissue a 3D warp knitted structure is required, and for this application, a warp knitted spacer fabric is the preferred structure. Spacer fabrics are defined as a 3 layer sandwich structure with two outer layers of fabric, each knitted on its own row of needles, and a third inner spacer layer as shown in **Figure 13A**. In addition to having the advantages of a warp knitted fabric, such as high bursting strength, high elongation, high porosity and low Young's modulus, the 3D spacer fabric is a one piece multi-layered structure with high volume to weight ratio, softness, breathability, moisture permeability, compression resistance and excellent recovery properties [65, 66].

The spacer yarn that lies in the thickness direction provides the mechanical support and the high total porosity needed for a tissue engineering scaffold. The yarns ensure

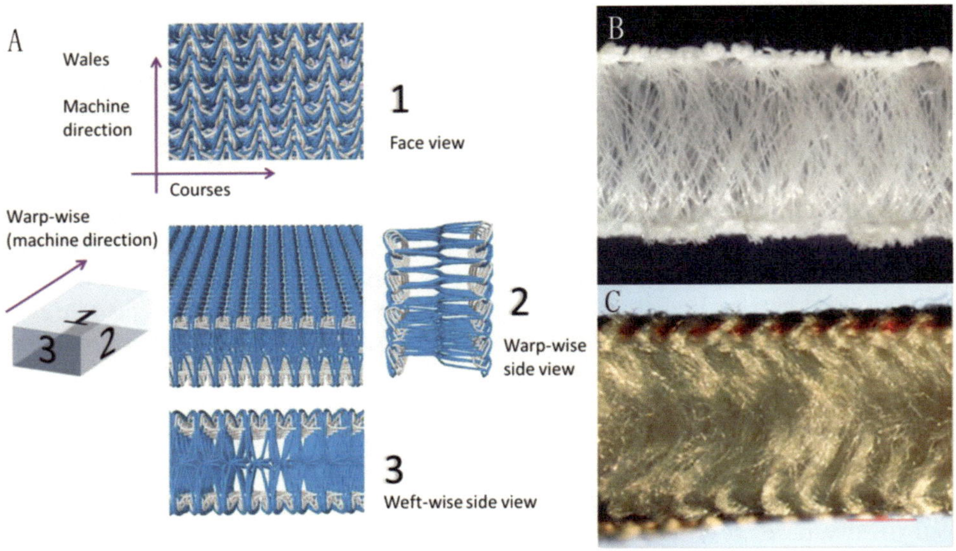

Figure 13.
ProCAD warp knit simulation of a 3D warp knitted spacer fabric (A); a spacer fabric knitted with monofilament yarns (B) and multifilament yarn (C) in the middle spacer layer [67].

a high surface area for cell attachment and proliferation, and the porous structure is highly interconnected which allow fluids carrying nutrients and waste by-products to flow through the entire structure [67]. The technical face, back and spacer layers are all knitted independently. So distinct characteristics can be designed and incorporated into the same fabric by means of changing the yarns and the construction in each layer. The distance between the two needle beds, also known as the void volume or "total porosity", can be altered by the knitting pattern in the spacer layer, which defines the macro level. At the same time, the size, twist and texture of the filament yarns, and their individual cross-sectional shape defines the micro level. The unique spacer layer determines the thickness of the scaffold, ranging from about 100 μm to several centimeters. Space fabrics are knitted from monofilament or multifilament yarns as shown in **Figure 13B** and **C**. A monofilament yarn consists of one thick filament per yarn so the yarn stiffness is higher and the fabric compression resistance is higher than a multifilament spacer fabric. Multi-filament yarns have several filaments per yarn, sometimes there can be hundreds of filaments. The higher the filament count, the finer each fiber. So multifilament spacer fabrics have extraordinary surface areas with high porosity [61].

Warp knitting is a promising technology to fabricate basic and complex scaffolds for tissue engineering applications. Warp knitted scaffolds have a high potential for commercial success because they can adapt to FDA-compliant materials without sacrificing their property requirements. The warp knitted fabric has great structural stability and suture retention performance, which is crucial in any clinical application. The unique type of 3D warp knitted spacer fabric, has proven to be biocompatible in lab trials with excellent cell attachment and tissue penetration into the 3D scaffold network. It is also an attractive candidate in complex tissue engineering applications such as at muscle-tendon junctions.

4.4 Braided scaffolds

Braiding technology, developed in the 1800s [68], is the process of interlacing three or more yarns obliquely to form either tubular or flat fabrics. In order to braid a tube one needs to use an even number of yarns, half rotating clockwise, and the other half rotating counter-clockwise. If a flat braid is needed, then one needs to use an odd number of ends. By increasing the number of sets of yarns or the thickness of the yarns, one can obtain a thicker or wider product with superior mechanical performance [69]. These designs can be a hollow or solid construction with either a uniform or variable cross-sectional shape [69]. Braiding technology has traditionally been used to fabricate textile structures such as ropes, but it is gradually gaining attention in biomedical applications, such as sutures, stents and in tissue engineering (TE) scaffolds for the repair of connective tissues, nerve guides and vascular prostheses.

4.4.1 Design of braided structure

Braiding angle, the most important geometrical parameter of braided structures, is defined as the angle between the braided yarns and the longitudinal direction. Braiding angles can range from 0° to 90° while they are usually between 30° and 80°. In comparison with a woven structure, the yarns in a braid are able to rotate and slide at the crossing points when under an external force. This offers braided structures with superior flexibility [70].

Tubular braids or ropes are mainly used in medical applications, which are manufactured with an even number of yarns arranged around a circle. The two common structures that have been developed are the diamond braided structure with each yarn crossing above and below the other yarns (1/1) and the regular braided structure with each yarn crossing over two of the other yarns (2/2) (**Figure 14**). In addition, other braided structures can be developed. For example, axial yarns can be introduced in the longitudinal direction to form a tri-axial braided structure [71].

4.4.2 Fabrication of braided scaffolds

The most widely used braiding machine for fabricating tubular braids has two sets of bobbins mounted on spindles moving along two tracks. One set of bobbins revolves in the clockwise direction and the other set revolves in the counter-clockwise direction, in order to form the braided pattern (**Figure 15**) [68, 72].

Yarns wrapped on the bobbins are pulled continuously and interlaced with each other at the braiding point. The braided yarns are then wound up on a scroll or take-up package. By adjusting the ratio of the braiding velocity and the take-up velocity, different braiding angles can be obtained which significantly affect the performance of the braided product.

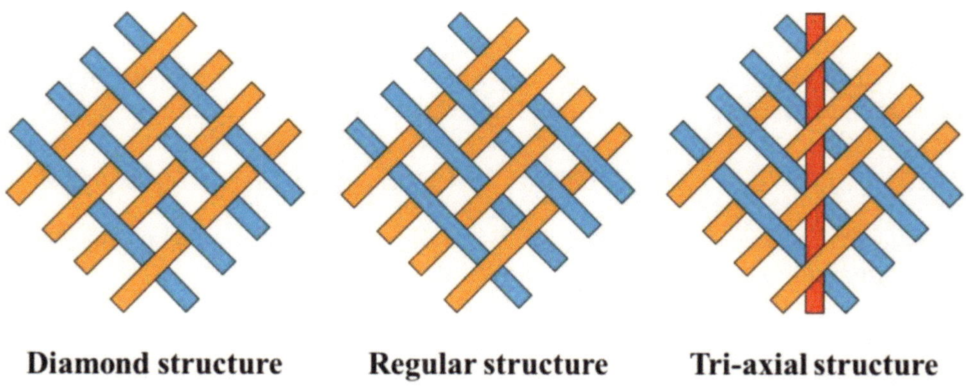

Figure 14.
Three types of braided structure.

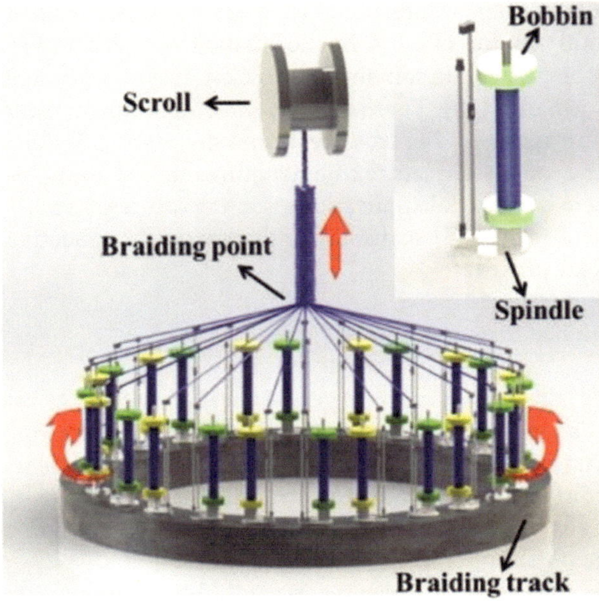

Figure 15.
Schematic diagram of braiding machine.

4.4.3 Evaluation of braided scaffolds

Braided scaffolds have frequently been used for tubular or rope-like tissue engineering scaffolds because of their precise and predictable porous structure and adjustable performance to mimic natural structures and properties [73, 74]. Extensive tissue ingrowth and mechanical characteristics that match natural tissues can be achieved with the appropriate selection of materials, braiding parameters and suitable pore size [75]. First, the mechanical and physical performance are considered as the most important criteria for designing a braided tissue engineering scaffold, followed by the question of biocompatibility.

4.4.3.1 Physical properties of braided scaffold

For rope-like braided scaffolds, the braiding angle is the key to affect their mechanical properties. The cover factor, combined with the braiding angle, needs to be measured in order to evaluate the physical properties of the tubular braided scaffold. The easiest way to measure braiding angle (θ) is by using a microscope and angle measurement software, as shown in **Figure 16**.

For tubular braided scaffolds, the cover factor is defined as the percentage of the mandrel's surface covered by yarns and calculated with Eq. (1) [76]:

$$\textit{Cover factor:} = 1 - \left(1 - \frac{W_y\ N_c}{4\pi R \cos\theta}\right)^2 \quad (1)$$

where W_y is the width of braiding yarns; N_c represents the number of bobbins; R is the outer radius of mandrel (mm) and θ is the braiding angle.

4.4.3.2 Mechanical properties of braided scaffold

Radial compression, tensile and bending measurements are commonly used to evaluate the mechanical properties of braided scaffolds, especially tubular braided scaffolds.

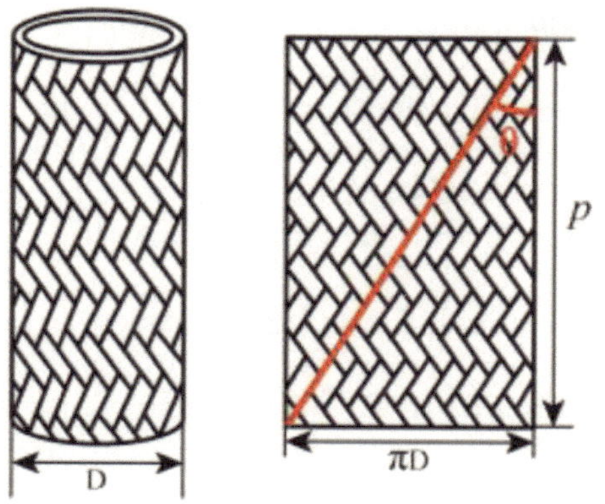

Figure 16.
Diagram of braiding angle. D: outer diameter of braided scaffold; p: axial distance of braiding yarn in one spiral.

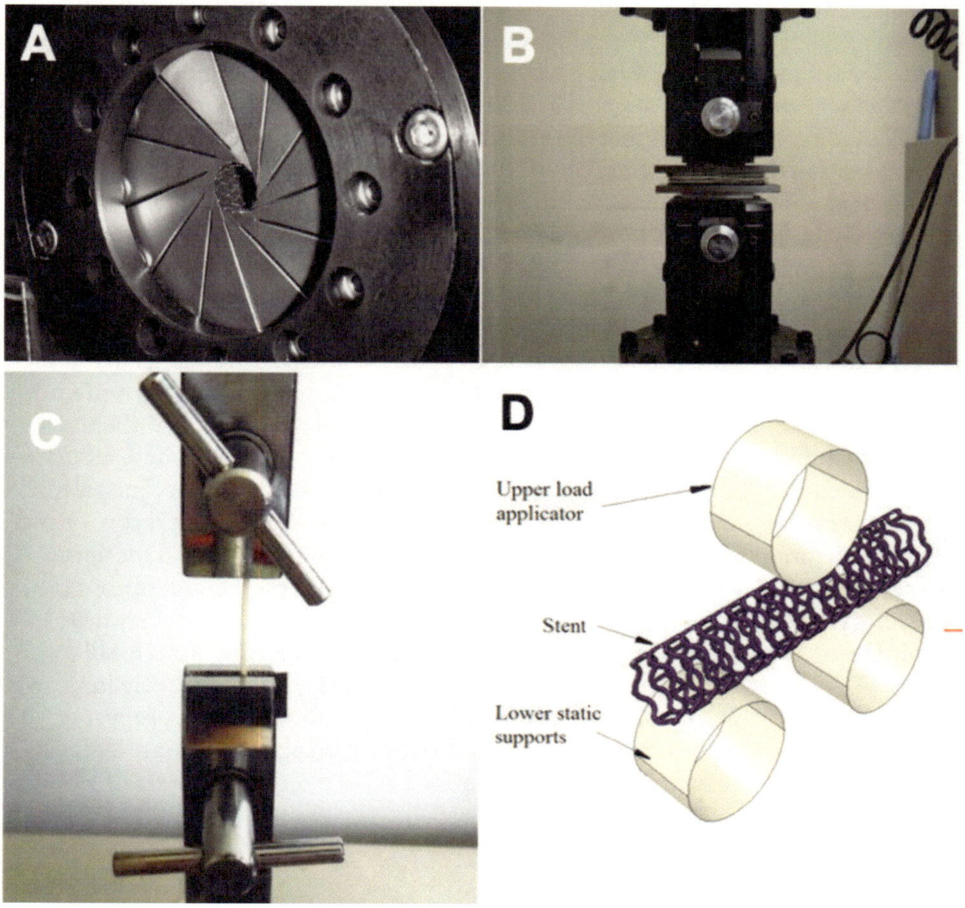

Figure 17.
(A) Radial compression machine; (B) parallel plate compression tester; (C) uniaxial tensile measurement; (D) diagram of three-point bending measurement.

According to ISO 25539-2012 [77], the radial force test with a radial compression machine and the crush resistance test with a parallel plate tester are highly recommended to evaluate the radial compressional properties of tubular braided scaffolds (**Figure 17A** and **B**) [78, 79]. A uniaxial tensile test (**Figure 17C**) [75, 80] and a three-point bending test (**Figure 17D**) [81] are measurements used to evaluate the tensile strength and bending stiffness of braided scaffolds, respectively.

4.4.4 Tissue engineering applications of braided scaffolds

For tissue engineering applications, braided scaffolds are required to have good cell adhesion and cell proliferation. At present, different biomaterials are used to fabricate various braided scaffolds for tissue engineering applications such as tendon/ligament reconstruction, cartilage, bone, vascular grafts and nerve regeneration.

Barber et al. [82] and Rothrauff et al. [83] reported that they had braided nanofibrous scaffolds (BNFSs) for tissue engineering tendons and ligaments. Several bundles of electrospun nanofibers were braided into rope-like scaffolds. Human mesenchymal stem cells (hMSCs) showed good adhesion and orientation after seeding and culturing on the BNFSs, and were also reported to promote hMSCs proliferation and key pluripotency gene expression. Cooper et al. [84] and Freeman et al. [85] developed braided scaffolds made with synthetic poly(L-lactic acid) for anterior cruciate ligament repair. The scaffolds mimicked the morphology and mechanical properties of the native ligament tissue and when tested in rabbits showed excellent healing and regeneration. In another study, Fang et al. developed a braided scaffold using antheraea pernyi silk fibroin for tissue engineering a tendon [86]. The scaffold was investigated *in vitro* and *in vivo* using tenocytes and a rabbit animal model. It was found that the scaffolds integrated with the native tissue and formed tendon tissue in rabbits.

Sun et al. fabricated a gene-modified scaffold by lyophilizing the CHS mixture with braided silk cables for fibrocartilage application [87]. The scaffold, seeded with mesenchymal stem cells (MSCs), showed vigorous cell proliferation and differentiation to reconstruct the cartilage. Fujihara et al. successfully developed a braided carbon/PEEK composite to be used as a bone plate [88]. The bending properties of the braided composite were comparable with natural bone. In another study, Evans et al. developed a tubular braid to improve the efficiency of bone fracture treatment [89]. It was shown that changing the braid angle and the thickness of the tubular cast produced a stiffness similar to that of native bone. Ichihara et al. braided PLLA and PGA yarns coated with collagen to form a novel nerve guide tube for nerve regeneration [90]. The animal experiment showed fast recovery and good regeneration by using the scaffold, which suggested the potential for nerve gap repair. Zhang et al. designed a tri-layer graft from electrospun silk fibroin (SF) and poly(L-lactide-co-ε-caprolactone) (PLCL) and braided layers of silk yarns, to mimic the tri-layer structure of the intima, media, and adventitia of native blood vessels [91]. It was demonstrated that the braided outer layer significantly improved the mechanical properties of the construct. Besides good mechanical properties, and biocompatibility the prototype sample also exhibited appropriate anticoagulation properties as a result of the heparin coating.

4.5 Nonwoven scaffolds

Nonwoven textiles can be distinguished from traditional textiles based on the fact that nonwoven textiles are manufactured directly from staple fibers or filaments and do not involve the intermediate yarn manufacturing process [92].

Various definitions of nonwoven textiles are used by various nonwoven organizations. One of the most widely used definition is the one defined by the Association of Nonwoven Fabrics Industry (INDA), which is "A sheet, web, or batt of natural and/or man-made fibers or filaments, excluding paper, that have not been converted into yarns, and that are bonded to each other by any of several means" [93].

Thus, nonwoven textiles essentially are characterized by those fabrics which are converted directly from fibers. As opposed to conventional textile manufacturing of woven or knitted fabrics, nonwoven manufacturing processes are characterized by cost effective and high productivity due to elimination of the yarn manufacturing step. The ability to process a wide range of raw materials from staple fibers to continuous polymer filaments adds to a versatility in the range of products that can be obtained from nonwoven textile processing [94].

4.5.1 Nonwoven fabrication process

Nonwoven fabric manufacturing consists of three main steps: (1) selection of raw material fiber or polymer, (2) web formation, and (3) web consolidation and finishing [95]. Selection of the raw material or type of fiber is generally based on the requirement of specific properties for the end use application. Raw material for nonwoven fabrics includes staple fibers or polymers, binders for thermal or chemical web consolidation and finishing agents such as softeners, flame retardants, antimicrobials [95]. On the basis of web formation, nonwovens can be classified into two main categories, web formation from staple fiber and web formation from polymers. Staple-fiber-based web formation is further divided into two categories: dry laid webs and wet laid webs.

Dry laid web formation includes opening and mixing of staple fibers and the formation of a thin layer of web of randomly laid or oriented individual fibers by air laying or a conventional carding machine. The single layered web is then laid into its final web structure by going through cross lapping or parallel laying to get a stable structural integrity. The orientation of fibers in the machine or cross direction can be engineered through these processes, depending upon the properties required in the final product [95]. Wet laid web formation includes initial mixing of the staple fibers in chemicals and water and then deposition of the mixture into a thin layer of web consisting of randomly laid fibers. The advantage of wet laid web formation is that fibers with very short staple length can be easily converted into a web [95]. Polymer based web formation is further divided into spunbonded and melt blown web formation. Both the processes include melting of the polymer into a dope which is then spun into fine filaments or microfibers and directly collected on a collection plate in the form of a web which can then be further processed for producing a hybrid nonwoven structure or can be finished into a final product [95]. Variables such as fiber type, fiber processing, type of web formation, web weight, uniformity and the presence of binder can impact the characteristics and properties of the final fabric [94].

Web consolidation or web bonding processes are classified into three main types, which are mechanical, chemical and thermal bonding. Different types of web consolidation processes impart specific properties to the nonwoven fabric. Mechanical web bonding consists of two types: needle punching and hydroentangling. In both the processes a randomly laid or carded web goes through penetration with either barbed needles or high-speed water jets resulting in interlocking and bonding of the web in the form of a fabric with strength and stability. The properties of the scaffold such as thickness, total porosity, air permeability can be engineered by various machine variables such as the intensity and depth of penetration of the needles, the number of entanglements per specific surface area [94]. In chemical bonding, the adhesion of fibers in the web is achieved with the help of binders such as polyvinyl acid derived resins. The characteristics and amount of binder used determines the properties of the nonwoven fabric (**Figure 18**).

For example, increasing the amount of binder results in increased strength and stiffness, but reduced softness and flexibility. Chemical bonding can be achieved by various methods such as saturation, foam, spray, and print bonding [94]. Thermal bonding involves binding the web using thermoplastic binders, such as fibers or powders. Thermal bonding techniques include hot calendering, air thermal

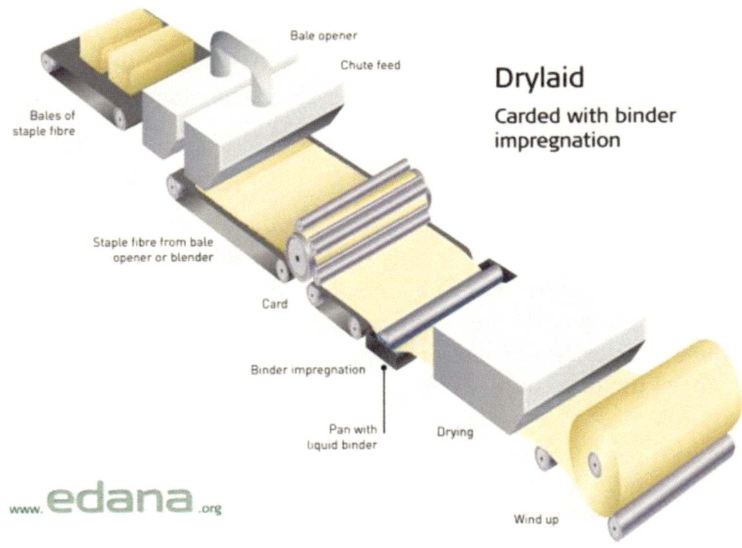

Figure 18.
Schematic diagram of nonwoven web formation using a carded drylaid web and bonding with liquid chemical binder impregnation and drying.

bonding, ultrasonic, and radiant heat bonding. In thermal bonding, various surface characteristics can be introduced using calender bonding of different profiles including area bonding, point bonding, embossing and grid bonding [94].

Finishing of nonwovens includes various mechanical and chemical finishes such as calendering, heat setting, shearing, singeing, and applying antimicrobial and antistatic agents. The chemical finishes can be applied through processes such as padding, coating, laminating and newer finishing techniques such as plasma and microencapsulation [94].

4.5.2 Nonwoven fabrics as scaffolds

Nonwoven fabrics are characterized by properties such as high surface area and high porosity, which encourages researchers to study various types of nonwoven fabrics for applications in tissue engineering. Though currently there is no commercial product available, various studies have been conducted on nonwoven fabrics as tissue engineering scaffolds for *in vitro* tissue regeneration [96–102]. Needle punched nonwoven scaffolds have been studied for cell attachment and proliferation of cell types including mesenchymal stromal cells, mouse fibroblasts and hepatocytes using different types of fibers such as polyvinylidene fluoride (PVDF), polyester, polypropylene and nylon fibers [96–98]. Polyester melt blown nonwoven scaffolds have also been studied for osteogenic differentiation using mesenchymal stem cells [99]. Another study incorporating nonwoven scaffolds included poly(vinyl alcohol) (PVA) bonded to a wet-laid chitosan nonwoven scaffold characterized by high porosity and a narrow pore size distribution. These scaffold properties are recommended for soft tissue regeneration, such as cartilage tissue [100]. Polymer impregnated Lyocell fiber nonwoven biomaterials were seeded with chondrocytes to study their ability to regenerate cartilage tissue [101]. Conductive scaffolds have also been fabricated by coating poly-L-lactide (PLLA) spunbond biomaterials with conductive poly(3,4-ethylenedioxythiophene) (PEDOT), and human dermal fibroblasts have been cultured to evaluate on the biocompatibility of the coated scaffold [102].

4.6 Electrospun scaffolds

Electrospinning is one of the most popular techniques for fabricating tissue engineering scaffolds. Its widespread use is accredited to its ease of manipulation, cheap and accessible equipment needs, and its versatility. The technique can be applied to various materials, ranging from synthetic polymers such as PLA [103], PGA [104], PCL [105, 106], PU [107, 108], and their copolymers [109], to natural polymers such as collagen [110, 111], elastin [112], gelatin [113] and chitosan [114]. Electrospun scaffolds have been applied in various tissue engineering applications, such as skin [115], bone [107, 116], cartilage [113, 117], tendon [118, 119], ligament [118], nerve [105, 120], blood vessel [121], cardiac tissue [122], and aortic valve [108].

Electrospinning is a fiber-forming method by injecting a conductive polymer solution or melt through a high-voltage field, and the fiber is stretched by the attenuating electrical force and collected on a grounded collector (**Figure 19**). It has a unique advantage in fabricating micro and nano-fibers that mimic the structure of the extra cellular matrix (ECM), such as the basement membrane of blood vessels. Hackett et al. [105], Zhang et al. [91] and our studies [57] have all demonstrated that an electrospun layer attached to the luminal surface of a vascular scaffold is able to reduce the pore size and facilitate endothelial cell proliferation. The nanofiber web provides an ideal scaffold surface for endothelial cells that require a flat surface with nano-sized pores to form a monolayer on the vascular intima.

Another advantage of electrospinning is that it is able to adapt to a wide range of materials. In order to fabricate a scaffold that mimicks the native ECM both structurally and biochemically, one can apply natural polymers such as collagen, elastin, and cellular components to the scaffold's surface. Collagen, as the predominant protein in native ECM, has been reported by many researchers to be an attractive scaffold coating that leads to advanced cell adhesion, proliferation, and migration [123, 124]. Furthermore, the addition of cellular components to the electrospinning solution has also been described. Venugopal et al. [125] reported preparing an electrospun scaffold from a blend of gelatin and phytochemical components, such as hexadecanoic acid (HDA), octadecanoic acid (ODA) and

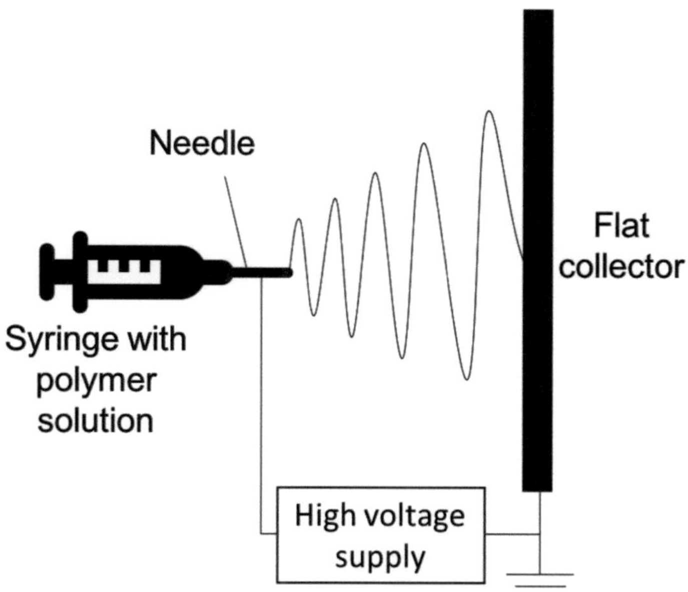

Figure 19.
Typical experimental electrospinning setup.

N,N-diisopropyl(2,2,3,3,3-pentafluoropropyl)amine (DPA), which is able to promote primary human meniscus cells and human MG63 osteoblast-like cells to attach and proliferate for bone and cartilage tissue regeneration.

On the other hand, its micro- or nanoscale structure turns out to be its limitation in fabricating a 3-dimentional scaffold. The small pore size prevents cell infiltration through its thickness, and so it does not regenerate bulk tissue with any thickness [108]. In addition, its mechanical weakness limits its translational ability. Thus, a composite scaffold by combining electrospinning with other textile technologies is an attractive strategy to take advantage of the different properties of different textile structures, and in this way promote the development of the novel and advanced tissue engineering scaffolds.

5. Conclusions

This chapter has demonstrated the use of a number of textile fiber-based technologies that can be used to prepare resorbable scaffolds for a wide variety of tissue engineering applications. Each textile fabrication technique has its particular advantages in being able to control its physical dimensions (e.g. average pore size and pore size distribution), its surface topography and its mechanical properties, whether they be related to supporting tensile, compression, bending and shear forces, within a precise and predictable range.

Few experimental prototype tissue engineering structures have been accepted clinically, and fewer still have been approved by the US regulatory agency for commercial production, distribution and use. Additional work is needed in order to understand the complex biomaterial-cell-tissue interactions that occur at the scaffold interface. It is hoped that by describing in this chapter the success of using fiber-based scaffolds that more efforts and collaborations among interdisciplinary research teams will be able to overcome these challenges.

Conflict of interest

None of the authors of this chapter have a 'conflict of interest' to declare.

Author details

Martin W. King[1,2*], Jiyang Chen[1], Monica Deshpande[1], Ting He[1,3], Harshini Ramakrishna[1], Yu Xie[1], Fan Zhang[1] and Fan Zhao[1,2]

1 North Carolina State University, Raleigh, USA

2 Donghua University, Shanghai, China

3 The Lycra Company, Wilmington, DE, USA

*Address all correspondence to: mwking2@ncsu.edu

References

[1] Langer R, Vacanti JP. Tissue engineering. Science. 1993;**260**:920-926. DOI: 10.1126/science.8493529

[2] Wolfe RA, Roys EC, Merion RM. Trends in organ donation and transplantation in the United States 1999-2008. American Journal of Transplantation. 2010;**10**:961-972

[3] MacArthur BD, Oreffo RO. Bridging the gap. Nature. 2005;**433**:19. DOI: 10.1038/433019a

[4] NSF Report. The emergence of tissue engineering as a research field [Internet]. 2004. Available from: https://www.nsf.gov/pubs/2004/nsf0450/start.htm [Accessed: 2019-01-15]

[5] Sumanasinghe RD, Bernacki SH, Loboa EG. Osteogenic differentiation of human mesenchymal stems cells in collagen matrices: Effect of uniaxial cyclic tensile strain on bone morphogenetic protein mRNA expression. Tissue Engineering. 2006;**12**(12):3459-3465. DOI: 10.1089/ten.2006.12. 3459

[6] Levin M. Reprogramming cells and tissue patterning via bioelectrical pathways: Molecular mechanisms and biomedical opportunities. Systems Biology & Medicine. 2013;**5**(6):657-676

[7] Meng S, Zhang Z, Rouabhia M. Accelerated osteoblast mineral-ization on a conductive substrate by multiple electrical stimulation. Journal of Bone Mineral Metabolism. 2011;**29**(5):535-544

[8] Griffith LG, Swartz MA. Capturing complex 3D tissue physiology in vitro. Nature Reviews: Molecular Cell Biology. 2006;**7**(3):211-224. DOI: 10.1038/nrm1858

[9] Horrocks AR, Anand SC. Handbook of Technical Textiles. Cambridge, UK: Woodhead Publishing; 2000

[10] Liberski A, Ayad N, Wojciechowska D, Kot R, Vo DM, Aibibu D, et al. Weaving for heart valve tissue engineering. Biotechnology Advances. 2017;**35**(6):633-656

[11] Clark JM. Tendons, ligaments, and capsule of the rotator cuff. Gross and microscopic anatomy. The Journal of Bone and Joint Surgery. American Volume. 1992;**74**(5):713-725

[12] Rothman RH, Parke WW. The vascular anatomy of the rotator cuff. Clinical Orthopaedics and Related Research. 1965;**41**:176-188

[13] Rathbun JB, Macnab I. The microvascular pattern of the rotator cuff. The Journal of Bone and Joint Surgery, British Volume. 1970;**52**(3):540-553

[14] Derwin KA, Codsi MJ, Milks RA, Baker AR, McCarron JA, Iannotti JP. Rotator cuff repair augmentation in a canine model with use of a woven poly-L-lactide device. The Journal of Bone and Joint Surgery. American Volume. 2009;**91**(5):1159-1171

[15] Proctor CS. Long-term successful arthroscopic repair of large and massive rotator cuff tears with a functional and degradable reinforcement device. Journal of Shoulder and Elbow Surgery. 2014;**23**(10):1508-1513

[16] Ratcliffe A, Butler DL, Dyment NA, Cagle PJ, Proctor CS, Ratcliffe SS, et al. Scaffolds for tendon and ligament repair and regeneration. Annals of Biomedical Engineering. 2015;**43**(3):819-831

[17] Hakimi O, Mouthuy PA, Zargar N, Lostis E, Morrey M, Carr A. A layered electrospun and woven surgical scaffold to enhance endogenous tendon repair. Acta Biomaterialia. 2015;**26**:124-135

[18] Wu S, Wang Y, Streubel PN, Duan B. Living nanofiber yarn-based woven

biotextiles for tendon tissue engineering using cell tri-culture and mechanical stimulation. Acta Biomaterialia. 2017;**62**:102-115

[19] Islam A, Bohl MS, Tsai AG, Younesi M, Gillespie R, Akkus O. Biomechanical evaluation of a novel suturing scheme for grafting load-bearing collagen scaffolds for rotator cuff repair. Clinical Biomechanics. 2015;**30**(7):669-675

[20] Younesi M, Islam A, Kishore V, Anderson JM, Akkus O. Tenogenic induction of human MSCs by anisotropically aligned collagen biotextiles. Advanced Functional Materials. 2014;**24**(36):5762-5770

[21] Gao Y, Shao W, Qian W, He J, Zhou Y, Qi K, et al. Biomineralized poly (l-lactic-co-glycolic acid)-tussah silk fibroin nanofiber fabric with hierarchical architecture as a scaffold for bone tissue engineering. Materials Science and Engineering: C. 2018;**84**:195-207

[22] Gilmore J, Burg T, Groff RE, Burg KJ. Design and optimization of a novel bio-loom to weave melt-spun absorbable polymers for bone tissue engineering. Journal of Biomedical Materials Research Part B: Applied Biomaterials. 2017;**105**(6):1342-1351

[23] Persson M, Lehenkari PP, Berglin L, Turunen S, Finnilä MA, Risteli J, et al. Osteogenic differentiation of human mesenchymal stem cells in a 3D woven scaffold. Scientific Reports. 2018;**8**(1):10457. DOI: 10.1038/s41598-018-28699-x

[24] Huynh NP, Brunger JM, Gloss CC, Moutos FT, Gersbach CA, Guilak F. Genetic engineering of mesenchymal stem cells for differential matrix deposition on 3D woven scaffold. Tissue Engineering. 2018;**24**(19-20):1531-1544

[25] Friedman JM, Sennett ML, Bonadio MB, Orji KO, Neuwirth AL, Keah N, et al. Comparison of fixation techniques of 3D-woven poly (ϵ-caprolactone) scaffolds for cartilage repair in a weight bearing porcine large animal model. Cartilage. 2017;**9**(4):428-437

[26] Moffat KL, Goon K, Moutos FT, Estes BT, Oswald SJ, Zhao X, et al. Composite cellularized structures created from an interpenetrating polymer network hydrogel reinforced by a 3D woven scaffold. Macromolecular Bioscience. 2018;**18**(10):1800140

[27] Baba S, Inoue T, Hashimoto Y, Kimura D, Ueda M, Sakai K, et al. Effectiveness of scaffolds with pre-seeded mesenchymal stem cells in bone regeneration—Assessment of osteogenic ability of scaffolds implanted under the periosteum of the cranial bone of rats. Dental Materials Journal. 2010;**29**(6):673-681

[28] Arjunon S, Rathan S, Jo H, Yoganathan AP. Aortic valve: Mechanical environment and mechanobiology. Annals of Biomedical Engineering. 2013;**41**(7):1331-1346

[29] Wu S, Duan B, Qin X, Butcher JT. Living nano-micro fibrous woven fabric/hydrogel composite scaffolds for heart valve engineering. Acta Biomaterialia. 2017;**51**:89-100

[30] Zhang L, Ao Q, Wang A, Lu G, Kong L, Gong Y, et al. A sandwich tubular scaffold derived from chitosan for blood vessel tissue engineering. Journal of Biomedical Materials Research Part A. 2006;**77**(2):277-284

[31] Soffer L, Wang X, Zhang X, Kluge J, Dorfmann L, Kaplan DL, et al. Silk-based electrospun tubular scaffolds for tissue-engineered vascular grafts. Journal of Biomaterials Science, Polymer Edition. 2008;**19**(5):653-664

[32] Wang F, Mohammed A, Li C, Ge P, Wang L, King MW. Degradable/non-degradable polymer composites

for in-situ tissue engineering small diameter vascular prosthesis application. Biomedical Materials and Engineering. 2014;**24**(6):2127-2133

[33] Xie Y, Guan Y, Kim SH, King MW. The mechanical performance of weft-knitted/electrospun bilayer small diameter vascular prostheses. Journal of Mechanical Behavior of Biomedical Materials. 2016;**61**:410-418

[34] Li C, Wang F, Douglas G, Zhang Z, Guidoin R, Wang L. Comprehensive mechanical characterization of PLA fabric combined with PCL to form a composite structure vascular graft. Journal of Mechanical Behavior of Biomedical Materials. 2017;**69**:39-49

[35] Li G, Liu Y, Lan P, Li Y, Li Y. A prospective bifurcated biomedical stent with seamless woven structure. Journal of the Textile Institute. 2013;**104**(9):1017-1023

[36] Yokota T, Ichikawa H, Matsumiya G, Kuratani T, Sakaguchi T, Iwai S, et al. In situ tissue regeneration using a novel tissue-engineered, small-caliber vascular graft without cell seeding. Journal of Thoracic and Cardiovascular Surgery. 2008;**136**(4):900-907

[37] Takahashi H, Yokota T, Uchimura E, Miyagawa S, Ota T, Torikai K, et al. Newly developed tissue-engineered material for reconstruction of vascular wall without cell seeding. Annals of Thoracic Surgery. 2009;**88**(4):1269-1276

[38] Iwai S, Sawa Y, Taketani S, Torikai K, Hirakawa K, Matsuda H. Novel tissue-engineered biodegradable material for reconstruction of vascular wall. Annals of Thoracic Surgery. 2005;**80**(5):1821-1827

[39] Bruer SM, Powell N, Smith G. Three-dimensionally knit spacer fabrics: A review of production techniques and applications. Journal of Textile and Apparel, Technology and Management. 2005;**4**(4):1-31

[40] Ramakrishna S. Textile scaffolds in tissue engineering. In: Tao XM, editor. Smart Fibres, Fabrics and Clothing. Boca Raton, FL, USA: CRC Press; 2001. pp. 291-313

[41] Wang X, Han C, Hu X, Sun H, You C, Gao C, et al. Applications of knitted mesh fabrication techniques to scaffolds for tissue engineering and regenerative medicine. Journal of the Mechanical Behavior of Biomedical Materials. 2011;**4**(7):922-932

[42] Doser M, Planck H. Textiles for implants and regenerative medicine. In: Bartels VT, editor. Handbook of Medical Textiles. Cambridge, UK: Woodhead Publishing; 2011. pp. 132-152. DOI: 10.1533/9780857093691.1.132

[43] Damme HV, Deprez M, Creemers E, Limet R. Intrinsic structural failure of polyester (Dacron) vascular grafts. A general review. Acta Chirurgica Belgica. 2005;**105**(3):249-255

[44] Nelson DM, Ma Z, Fujimoto KL, Hashizume R, Wagner WR. Intra-myocardial biomaterial injection therapy in the treatment of heart failure: Materials, outcomes and challenges. Acta Biomaterialia. 2011;**7**(1):1-5

[45] Fujimoto KL, Guan J, Oshima H, Sakai T, Wagner WR. In vivo evaluation of a porous, elastic, biodegradable patch for reconstructive cardiac procedures. Annals of Thoracic Surgery. 2007;**83**(2):648-654

[46] Boublik J, Park H, Radisic M, Tognana E, Chen F, Pei M, et al. Mechanical properties and remodeling of hybrid cardiac constructs made from heart cells, fibrin, and biodegradable, elastomeric knitted fabric. Tissue Engineering. 2005;**11**(7-8):1122-1132

[47] Chen J, Cheng K, West AJ, King MW. A biodegradable knitted cardiac patch for myocardium regeneration using cardiosphere-derived cells (CDCs).

Journal of Donghua University (English Edition). 2017;**2**:028

[48] Kitahara M, Miyagawa S, Fukushima S, Saito A, Shintani A, Akita T, et al. Biodegradable vs nonbiodegradable cardiac support device for treating ischemic cardiomyopathy in a canine heart. Seminars in Thoracic and Cardiovascular Surgery. 2017;**29**(1):51-61

[49] Starling RC, Jessup M, Oh JK, Sabbah HN, Acker MA, Mann DL, et al. Sustained benefits of the CorCap cardiac support device on left ventricular remodeling: Three year follow-up results from the Acorn clinical trial. Annals of Thoracic Surgery. 2007;**84**(4):1236-1242

[50] Gilard M, Eltchaninoff H, Iung B, Donzeau-Gouge P, Chevreul K, Fajadet J, et al. Registry of transcatheter aortic-valve implantation in high-risk patients. New England Journal of Medicine. 2012;**366**(18):1705-1715

[51] Kodali SK, Williams MR, Smith CR, Svensson LG, Webb JG, Makkar RR, et al. Two-year outcomes after transcatheter or surgical aortic-valve replacement. New England Journal of Medicine. 2012;**366**(18):1686-1695

[52] Mack MJ, Leon MB, Smith CR, Miller DC, Moses JW, Tuzcu EM, et al. Five year outcomes of transcatheter aortic valve replacement or surgical aortic valve replacement for high surgical risk patients with aortic stenosis (PARTNER 1): A randomised controlled trial. The Lancet. 2015;**385**(9986):2477-2484

[53] Lieshout MV, Peters G, Rutten M, Baaijens F. A knitted, fibrin-covered polycaprolactone scaffold for tissue engineering of the aortic valve. Tissue Engineering. 2006;**12**(3):481-487

[54] Amat-Santos IJ, Dahou A, Webb J, Dvir D, Dumesnil JG, Allende R, et al. Comparison of hemodynamic performance of the balloon-expandable SAPIEN 3 versus SAPIEN XT transcatheter valve. American Journal of Cardiology. 2014;**114**(7):1075-1082

[55] Faxon DP, Fuster V, Libby P, Beckman JA, Hiatt WR, Thompson RW, et al. Atherosclerotic vascular disease conference: Writing group III: Pathophysiology. Circulation. 2004;**109**(21):2617-2625

[56] Cong H, Xie Y, Chapin K, Learn G, Akkus O, King M. Characterization of aligned collagen yarns for biomedical textile fabrication. Frontiers in Bioengineering and Biotechnology. 2016. DOI: 10.3389/conf.FBIOE.2016.01.02502

[57] Zhang F, Xie Y, Celik H, Akkus O, Bernacki SH, King MW. Engineering small-caliber vascular grafts from collagen filaments and nanofibers with comparable mechanical properties to native vessels. Biofabrication. 2019. DOI: 10.1088/1758-5090/ab15ce

[58] Naghavi Alhosseini S, Moztarzadeh F, Kargozar S, Dodel M, Tahriri M. Development of polyvinyl alcohol fibrous biodegradable scaffolds for nerve tissue engineering applications: In vitro study. International Journal of Polymeric Materials and Polymeric Biomaterials. 2015;**64**(9):474-480

[59] Li G, Li Y, Chen G, He J, Han Y, Wang X, et al. Silk-based biomaterials in biomedical textiles and fiber-based implants. Advanced Healthcare Materials. 2015;**4**(8):1134-1151

[60] Wang A, Ao Q, Cao W, Yu M, He Q, Kong L, et al. Porous chitosan tubular scaffolds with knitted outer wall and controllable inner structure for nerve tissue engineering. Journal of Biomedical Materials Research Part A. 2006;**79**(1):36-46

[61] Ting H. A study of three dimensional warp knits for novel applications as tissue engineering scaffolds [MS thesis]. Raleigh: North Carolina State University; 2011

[62] Application of DynaMesh-CICAT [Internet]. 2011. Available from: en.dyna-mesh.com [Accessed: 2019-01-15]

[63] Elsner P, Hatch KL, Wigger-Alberti W, editors. Textiles and the Skin. New York: Karger; 2003

[64] Peter G, Biomaterials unlock potential of regenerative medicine [Internet]. 2012. Available from: https://www.medicaldesignbriefs.com/component/content/article/mdb/features/14341 [Accessed: 2019-01-15]

[65] Rajendran S, Anand S. Design and development of novel bandages for compression therapy. British Journal of Nursing. 2003;**12**(Sup1):S20-S29

[66] Miao XH, Ge MQ. The compression behavior of warp knitted spacer fabric. Fibres and Textiles in Eastern Europe. 2008;**16**(1):90-92

[67] Ting H. Tissue engineering 3D textiles scaffolds for multiple junction tissue regeneration [PhD thesis]. Raleigh: North Carolina State University; 2016

[68] Li C, Wang X, Zhao F, Wang F, Ting Z, Wang L. Modeling of braiding parameter impact on pore size and porosity in a tubular braiding fabric. Polymers. 2017;**17**(3):221-226. DOI: 10.1515/epoly-2016-0060

[69] Kyosev Y. Braiding technology for textiles: Principles, design and processes. Cambridge, UK: Woodhead Publishing; 2014

[70] Ko FK. Tensile strength and modulus of a three-dimensional braid composite. In: Whitney J, editor. Composite Materials: Testing and Design. STP892-EB. West Conshohocken, PA, USA: ASTM International; 1986. pp. 392-403

[71] Alpyildiz T. 3D geometrical modelling of tubular braids. Textile Research Journal. 2012;**82**(5):443-453. DOI: 10.1177/0040517511427969

[72] Du GW, Ko FK. Unit cell geometry of 3-D braided structures. Journal of Reinforced Plastics and Composites. 1993;**12**(7):752-768

[73] Puppi D, Zhang X, Yang L, Chiellini F, Sun X, Chiellini E. Nano/microfibrous polymeric constructs loaded with bioactive agents and designed for tissue engineering applications: A review. Journal of Biomedical Materials Research Part B: Applied Biomaterials. 2014;**102**(7): 1562-1579. DOI: 10.1002/jbm.b.33144

[74] Aibibu D, Hild M, Cherif C. An overview of braiding structure in medical textile: Fiber-based implants and tissue engineering. In: Kyosev Y, editor. Advances in Braiding Technology. Cambridge, UK: Woodhead Publishing; 2016. pp. 171-190. DOI: 10.1016/B978-0-08-100407-4.00006-5

[75] Cooper JA, Lu HH, Ko FK, Freeman JW, Laurencin CT. Fiber-based tissue-engineered scaffold for ligament replacement: Design considerations and *in vitro* evaluation. Biomaterials. 2005;**26**(13):1523-1532

[76] Rebelo R, Vila N, Fangueiro R, Carvalho S, Rana S. Influence of design parameters on the mechanical behavior and porosity of braided fibrous stents. Materials & Design. 2015;**86**:237-247. DOI: 10.1016/J.MATDES. 2015.07.051

[77] ISO 25539-2:2012—Cardiovascular implants—Endovascular devices—Part 2: Vascular stents 2012. Available from www.iso.org/standard/60604.html [Accessed: 2018-12-06]

[78] Isayama H, Nakai Y, Toyokawa Y, Togawa O, Gon C, Ito Y, et al. Measurement of radial and axial forces of biliary self-expandable metallic stents. Gastrointestinal Endoscopy. 2009;**70**(1):37-44. DOI: 10.1016/J.GIE.2008.09.032

[79] Tokuda T, Shomura Y, Tanigawa N, Kariya S, Komemushi A, Kojima H, et al. Mechanical characteristics of composite knitted stents. Cardiovascular and Interventional Radiology. 2009;**32**(5):1028-1032. DOI: 10.1007/s00270-009-9622-7

[80] Zou T, Wang L, Li W, Wang W, Chen F, King MW. A resorbable bicomponent braided ureteral stent with improved mechanical performance. Journal of the Mechanical Behavior of Biomedical Materials. 2014;**38**:17-25. DOI: 10.1016/J.JMBBM.2014.06.004

[81] Wang Q, Fang G, Zhao Y, Wang G, Cai T. Computational and experimental investigation into mechanical performances of poly-L-lactide acid (PLLA) coronary stents. Journal of the Mechanical Behavior of Biomedical Materials. 2017;**65**:415-427. DOI: 10.1016/J.JMBBM.2016.08.033

[82] Barber JG, Handorf AM, Allee TJ, Li WJ. Braided nanofibrous scaffold for tendon and ligament tissue engineering. Tissue Engineering Part A. 2011;**19**(11-12):1265-1274

[83] Rothrauff BB, Lauro BB, Yang G, Debski RE, Musahl V, Tuan RS. Braided and stacked electrospun nanofibrous scaffolds for tendon and ligament tissue engineering. Tissue Engineering Part A. 2017;**23**(9-10):378-389. DOI: 10.1089/ten.tea.2016.0319

[84] Cooper JA, Sahota JS, Gorum WJ, Carter J, Doty SB, Laurencin CT. Biomimetic tissue-engineered anterior cruciate ligament replacement. Proceedings of the National Academy of Sciences. 2007;**104**(9):3049-3054

[85] Freeman JW, Woods MD, Cromer DA, Wright LD, Laurencin CT. Tissue engineering of the anterior cruciate ligament: The viscoelastic behavior and cell viability of a novel braid–twist scaffold. Journal of Biomaterials Science, Polymer Edition. 2009;**20**(12):1709-1728. DOI: 10.1163/156856208X386282

[86] Fang Q, Chen D, Yang Z, Li M. In vitro and in vivo research on using Antheraea pernyi silk fibroin as tissue engineering tendon scaffolds. Materials Science and Engineering: C. 2009;**29**(5):1527-1534. DOI: 10.1016/J.MSEC.2008.12.007

[87] Sun L, Li H, Qu L, Zhu R, Fan X, Xue Y, et al. Immobilized lentivirus vector on chondroitin sulfate-hyaluronate acid-silk fibroin hybrid scaffold for tissue-engineered ligament-bone junction. BioMed Research International. 2014;**2014**(68):1-10. DOI: 10.1155/2014/816979

[88] Fujihara K, Huang ZM, Ramakrishna S, Satkunanantham K, Hamada H. Development of braided carbon/PEEK composite bone plates. Advanced Composite Letters. 2001;**10**(1):13-20

[89] Evans KR, Carey JP. Feasibility of a braided composite for orthopedic bone cast. Biomedical Engineering Journal. 2013;7:9-17. DOI: 10.2174/1874120701307010009

[90] Ichihara S, Inada Y, Nakada A, Endo K, Azuma T, Nakai R, et al. Development of new nerve guide tube for repair of long nerve defects. Tissue Engineering Part C: Methods. 2009;**15**(3):387-402. DOI: 10.1089/ten.tec.2008.0508

[91] Zhang Y, Li XS, Guex AG, Liu SS, Müller E, Malini RI, et al. A compliant and biomimetic three-layered vascular graft for small blood vessels. Biofabrication. 2017;**9**(2):025010. DOI: 10.1088/1758-5090/aa6bae

[92] Russell SJ, editor. Handbook of Nonwovens. Cambridge: Woodhead Publishing; 2007

[93] Jirsák O, Wadsworth LC. Nonwoven Textiles. Durham, NC: Carolina Academic Press; 1999

[94] Karthik T, Rathinamoorthy R. Nonwovens: Process, Structure, Properties and Applications. Milton: WPI Publishing; 2017

[95] Turbak AF, editor. Nonwovens: theory, process, performance, and testing. Las Vegas, NV, USA: ICON Group International; 1993

[96] Schellenberg A, Ross R, Abagnale G, Joussen S, Schuster P, Arshi A, et al. 3D non-woven polyvinylidene fluoride scaffolds: Fibre cross section and texturizing patterns have impact on growth of mesenchymal stromal cells. PLoS One. 2014;**9**(4):e94353. DOI: 10.1371/journal.pone.0094353

[97] Edwards SL, Russell SJ, Ingham E, Matthews JB, Mitchell W. Nonwoven scaffolds of improved design for the tissue engineering of the anterior cruciate ligament. In: Anand SC, Kennedy JF, Miraftab M, Rajendran S, editors. Medical Textiles and Biomaterials for Healthcare. Boca Raton, FL, USA: CRC Press; 2006. pp. 355-365. DOI: 10.1533/9781845694104.6.355

[98] Rossouw CL, Chetty A, Moolman FS, Birkholtz LM, Hoppe H, Mancama DT. Thermo-responsive non-woven scaffolds for "smart" 3D cell culture. Biotechnology and Bioengineering. 2012;**109**(8):2147-2158

[99] Takahashi Y, Tabata Y. Effect of the fiber diameter and porosity of non-woven PET fabrics on the osteogenic differentiation of mesenchymal stem cells. Journal of Biomaterials Science, Polymer Edition. 2004;**15**(1):41-57. DOI: 10.1163/156856204322752228

[100] Jou CH, Chen WC, Yang MC, Hwang MC, Chou WL, Lin SM, et al. In vitro biocompatibility of three-dimensional chitosan scaffolds immobilized with chondroitin-6-sulfate. Polymers for Advanced Technologies. 2008;**19**(5):377-384. DOI: 10.1002/pat.1020

[101] Müller FA, Müller L, Hofmann I, Greil P, Wenzel MM, Staudenmaier R. Cellulose-based scaffold materials for cartilage tissue engineering. Biomaterials. 2006;**27**(21):3955-3963. DOI: 10.1016/j.biomaterials.2006.02.031

[102] Niu X, Rouabhia M, Chiffot N, King MW, Zhang Z. An electrically conductive 3D scaffold based on a nonwoven web of poly(l-lactic acid) and conductive poly (3, 4-ethylenedioxythiophene). Journal of Biomedical Materials Research Part A. 2015;**103**(8):2635-2644. DOI: 10.1002/jbm.a. 35408

[103] Xu T, Yao Q, Miszuk JM, Sanyour HJ, Hong Z, Sun H, et al. Tailoring weight ratio of PCL/PLA in electrospun three-dimensional nanofibrous scaffolds and the effect on osteogenic differentiation of stem cells. Colloids and Surfaces B: Biointerfaces. 2018;**171**:31-39

[104] Hajiali H, Shahgasempour S, Naimi-Jamal MR, Peirovi H. Electrospun PGA/gelatin nanofibrous scaffolds and their potential application in vascular tissue engineering. International Journal of Nanomedicine. 2011;**6**:2133-2141

[105] Hackett JM, Dang TT, Tsai EC, Cao X. Electrospun biocomposite polycaprolactone/collagen tubes as scaffolds for neural stem cell differentiation. Materials. 2010;**3**(6):3714-3128

[106] Li B, Liu C, Zhou F, Mao X, Sun R. Preparation of electrospun core-sheath yarn with enhanced bioproperties for biomedical materials. Biotechnology Letters. 2018;**40**(2):279-284

[107] Chao CY, Mani MP, Jaganathan SK. Engineering electrospun multi-component polyurethane scaffolding platform comprising grapeseed oil and honey/propolis for bone tissue regeneration. PLoS One. 2018;**13**(10):e0205699

[108] Puperi DS, Kishan A, Punske ZE, Wu Y, Cosgriff-Hernandez E, West JL, et al. Electrospun polyurethane and hydrogel composite scaffolds as biomechanical mimics for aortic valve tissue engineering. Biomaterials Science & Engineering. 2016;**2**(9):1546-1558

[109] Brown JH, Das P, DiVito MD, Ivancic D, Tan LP, Wertheim JA. Nanofibrous PLGA electrospun scaffolds modified with type I collagen influence hepatocyte function and support viability in vitro. Acta Biomaterialia. 2018;**73**:217-227

[110] Baek J, Sovani S, Choi W, Jin S, Grogan SP, D'Lima DD. Meniscal tissue engineering using aligned collagen fibrous scaffolds: Comparison of different human cell sources. Tissue Engineering Part A. 2018;**24**(1-2):81-93

[111] Dong B, Arnoult O, Smith ME, Wnek GE. Electrospinning of collagen nanofiber scaffolds from benign solvents. Macromolecular Rapid Communications. 2009;**30**(7):539-542

[112] Buttafoco L, Kolkman NG, Engbers-Buijtenhuijs P, Poot AA, Dijkstra PJ, Vermes I, et al. Electrospinning of collagen and elastin for tissue engineering applications. Biomaterials. 2006;**27**(5):724-734

[113] Honarpardaz A, Irani S, Pezeshki-Modaress M, Zandi M, Sadeghi A. Enhanced chondrogenic differentiation of bone marrow mesenchymal stem cells on gelatin/glycosaminoglycan electrospun nanofibers with different amount of glycosaminoglycan. Journal of Biomedical Materials Research Part A. 2019;**107**(1):38-41

[114] Pezeshki-Modaress M, Zandi M, Rajabi S. Tailoring the gelatin/chitosan electrospun scaffold for application in skin tissue engineering: An in vitro study. Progress in Biomaterials. 2018;7(3):207-218

[115] Chummun I, Bhaw-Luximon A, Jhurry D. Modulating matrix-multicellular response using polysucrose-blended with poly-l-lactide or polydioxanone in electrospun scaffolds for skin tissue regeneration. Journal of Biomedical Materials Research Part A. 2018;**106**(12):3275-3291

[116] Zhang K, Wang Y, Sun T, Wang B, Zhang H. Bioinspired surface functionalization for improving osteogenesis of electrospun polycaprolactone nanofibers. Langmuir. 2018;**34**(50):15544-15550

[117] Jiang T, Kai D, Liu S, Huang X, Heng S, Zhao J, et al. Mechanically cartilage-mimicking poly(PCL-PTHF urethane)/collagen nanofibers induce chondrogenesis by blocking NF-kappa B signaling pathway. Biomaterials. 2018;**178**:281-292

[118] Wu G, Deng X, Song J, Chen F. Enhanced biological properties of biomimetic apatite fabricated polycaprolactone/chitosan nanofibrous bio-composite for tendon and ligament regeneration. Journal of Photochemistry and Photobiology B: Biology. 2018;**178**:27-32

[119] Orr SB, Chainani A, Hippensteel KJ, Kishan A, Gilchrist C, Garrigues NW, et al. Aligned multilayered electrospun scaffolds for rotator cuff tendon tissue engineering. Acta Biomaterialia. 2015;**24**:117-126

[120] Chang W, Shah MB, Lee P, Yu X. Tissue-engineered spiral nerve guidance conduit for peripheral nerve regeneration. Acta Biomaterialia. 2018;**73**:302-311

[121] Bertram U, Steiner D, Poppitz B, Dippold D, Köhn K, Beier JP, et al. Vascular tissue engineering: Effects of integrating collagen into a PCL based nanofiber material. BioMed Research International. 2017;**2017**:9616939

[122] Wang L, Wu Y, Hu T, Guo B, Ma PX. Electrospun conductive nanofibrous scaffolds for engineering cardiac tissue and 3D bioactuators. Acta Biomaterialia. 2017;**59**:68-81

[123] Yang X, Li Y, He W, Huang Q, Zhang R, Feng Q. Hydroxyapatite/collagen coating on PLGA electrospun fibers for osteogenic differentiation of bone marrow mesenchymal stem cells. Journal of Biomedical Materials Research Part A. 2018;**106**(11):2863-2870

[124] Joshi J, Brennan D, Beachley V, Kothapalli CR. Cardiomyogenic differentiation of human bone marrow-derived mesenchymal stem cell spheroids within electrospun collagen nanofiber mats. Journal of Biomedical Materials Research Part A. 2018;**106**(12):3303-3312

[125] Venugopal E, Rajeswaran N, Sahanand KS, Bhattacharyya A, Rajendran S. In vitro evaluation of phytochemical loaded electrospun gelatin nanofibers for application in bone and cartilage tissue engineering. Biomedical Materials. 2018;**14**(1):015004

Chapter 7

Microbioreactors and Perfusion Bioreactors for Microbial and Mammalian Cell Culture

Selvan Ravindran, Pooja Singh, Sanjay Nene, Vinay Rale, Nutan Mhetras and Anuradha Vaidya

Abstract

Screening for novel producer strains and enhanced therapeutic production at reduced cost has been the focus of most of the biopharmaceutical industries. The obligation to carry out prolonged intensive pilot scale experiments gave birth to micro-scale bioreactor systems. Screening large number of microorganisms using shake flasks and benchtop bioreactors is tedious and consumes resources. Microbioreactors that mimic benchtop bioreactors are capable not only of high throughput screening of producer strains, but also aid in optimizing the growth kinetics and expression of proteins. Modern technology has enabled the collection of precise online data for variables such as optical density (OD), pH, temperature, dissolved oxygen (DO), and adjusting in mixing inside microreactors. Microbioreactors have become an irreplaceable tool for biochemical engineers and biotechnologists to perform a large number of experiments simultaneously. Another aspect that is vital to any industry is the product yield and subsequent downstream processing. Perfusion bioreactors are one of the upcoming advances in bioreactor systems that have the potential to revolutionize biologics production. This chapter intends to take a review of different aspects of microbioreactors and perfusion bioreactors including their potential in high throughput pilot studies and microbial and mammalian cell cultivation technologies.

Keywords: microbioreactors, biopharmaceuticals, bioprocess, perfusion microbioreactors, microorganisms

1. Introduction

1.1 Need for microbioreactors in bioprocess development

Microbioreactor systems are an integral part of bioprocess engineering. In the past few years, microbioreactors have extensively been used for high throughput screening [1]. In addition to this, extensive bioprocess experiments were monitored and controlled. Industries involved in bulk production of pharmaceuticals, chemicals, enzymes for feed and food from microbial cell factories are in need of microbioreactors [2–4]. Advanced shaker microliter cultivation devices or down-scaled stirred tank reactors are two basic microbioreactors. Primary and secondary screening experiments based on microbial library are conducted in shake flasks and micro liter plates. This screening process aids in selecting the microbial strain candidates that are promising. Then, selected microbial candidates are subjected to lab-scale experimental conditions for better bioprocess control. Bioprocess method developed by microbioreactors is followed by successful lab-scale testing, and it is transferred to pilot scale. Pilot scale experiments are essential to understand the bioreactor inhomogeneities which can ultimately be addressed using simulators that can scale down the process of bioreactors [5, 6].

Traditional microbiorector experiments start from primary screening; followed by secondary screening using shake flasks; then, process development; and, finally, optimization leading to process validation and pilot scale. It is worth mentioning that microbioreactors reduce the number of steps involved in traditional bioreactors. Microbioreactors start with primary screening, followed by accelerated bioprocess development for secondary screening, resulting in process validation and pilot scale at a faster rate. Introduction of bioreactors has reduced the number of steps involved in traditional bioreactors to scale up the bioprocesses. Hence, microbioreactors have proved to me more economical as they reduce the usage of secondary screening by shake flasks, thus making the process development and optimization more efficient. Ecology and environmental concerns arising during traditional bioprocess development have also been addressed by microbioreactors. Therefore, the focus of this chapter is on new developments in microbioreactors and their impact on bioprocesses.

1.2 Advantages and expected outcome of microbioreactors

Microbioreactors have numerous advantages over traditional bioreactors during bioprocess development. Microbioreactors are necessary for high throughput quantitative microbial phenotyping under controlled experimental conditions. Microbioreactors reduce the time involved in the traditional bioprocess development by replacing the shake flasks and large lab-scale bioreactors (**Figure 1**). Microbioreactors are more than necessary with specifications for efficient and economical bioprocess development. Expected outcomes of microbioreactors are as follows: (a) reduction in volume for each cultivation experiment; (b) time conservation during experiments; (c) simple, friendly, and operational without

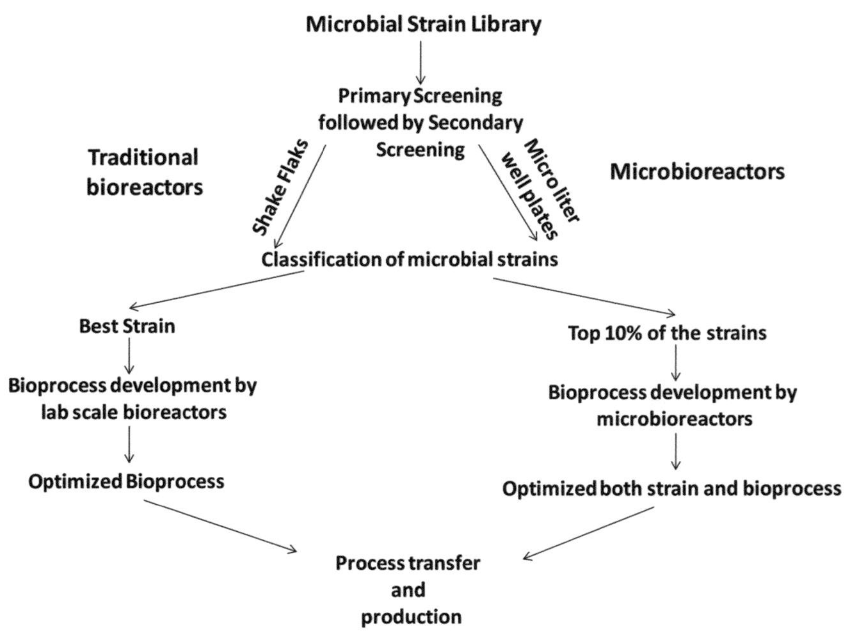

Figure 1.
Comparison of workflow for traditional bioreactors and microbioreactors.

fail; (d) automated operation with minimal or no supervision; (e) examination of bioprocess variables with high resolution; (f) temperature, pH, and feeding profiles for controlled cultivation process; (h) culture accessibility for sampling and dosing; (i) cultivation by fed-batch, batch, and continuous modes; (j) robotic systems with advanced hardware and software; and (k) scalability is compared to laboratory bioreactors [7].

For realizing all the above-mentioned desired properties of microbioreactors, researchers are emphasizing on applied and basic research. Extensive research has resulted in various microbioreactors with different technologies and applications.

2. Types of microbioreactors

Commercially available microbioreactors are based on their applications, specifications, and capabilities. Different types of microbioreactors are born out of extensive basic and applied research.

Biolector (m2p-labs.com), a commercial microbioreactor manufacturer, had developed microbioreactors for organism phenotyping, screening the strains, toxicity screening, and optimization of feed and growth parameters. These bioreactors support single-use 48-well plates that can hold culture volume of 0.8–2.4 ml. Biomass formation can be monitored via fluorescence and by optodes pH and dissolved oxygen (DO). This system is integrated with liquid handlers for pH adjustment; feeding; sampling; and control over temperature, gas, and humidity [8].

Microbioreactors (32 plate and 48 plate) [9] from Biolector were used for optimization of feeding rate, media screening, and fermentation parameters for anaerobic and microaerophyllic organisms. Few other applications that are possible with biolector microbioreactors are growth characterization, high throughput protein characterization, enzyme and cell activity tests, functional genomics and proteomic studies, inhibition and toxicity studies, and quality control.

RoboLector [10] is another microbioreactor where microbioreactor system from biolector is interfaced with liquid handling robot. These microbioreactors with the aid of 48 or 96 parallel cultivations through microplates provide fermentation data repeatedly for every 5–15 minutes. In addition to this, a robotic system also controls nutrient feeding, adjusts pH by adding acid or base, and does sampling based on user definition.

Another commercial manufacturer Micro-24 (pall.com) [11] developed a microbioreactor for screening strains and cell lines besides optimization of growth and feed parameters. These microbioreactors support single-use cassettes with 24 columns with culture volume ranging from 3 to 7 ml and regulate pH using ammonia, carbon dioxide, and control over temperature.

Microbioreactors to study biotransformation, phenotyping, strain and toxicity screening were developed by commercial manufacturer Bioscreen C (bioscreen.fi). These microbioreactors consist of two parallel 100-well plates with a capacity to hold culture volume of 0.4 ml. Optical density was used for quasi-continuous monitoring of biomass.

Growth profiler (enzyscreen.com), Applikon (applikonbio.com), SensorDish Reader (presens.de), ambr 15 (tapbiosystems.com), bioReactor (emag.de), and Sartorius (Sartorius.com) are other manufacturers of microbioreactors for various applications as per the requirements of clients.

2.1 Microbioreactors for fed-batch cultivation of *Escherichia coli*

During recombinant protein expression, *E. coli* synthesizes proteins at a faster rate with low multiplication rate. In traditional microbioreactors, fed-batch cultivations of *E. coli* were performed by stirred tank reactors [1]. Carbon source, magnesium, and ammonium were used as feeding solutions to match the nutritional requirements. It is worth mentioning that two fed-batch cultivations can be performed simultaneously with different nutrient compositions in the RoboLector systems. For example, feeding solution of one set of reaction is constituted with 1 M sodium hydroxide, 400 g/l glycerol, 100 g/l ammonium phosphate, and 1 g/l of magnesium sulfate $7H_2O$ whereas that of another set is composed of 200 g/l glycerol, 100 g/l ammonium phosphate, 2 g/l of magnesium sulfate $7H_2O$, and 1 M NaOH. Cultivations were performed at a shaking frequency of 1100 rpm with diameter 3 mm and temperature was set at 30°. Modified medium with 10 g/l glycerol and 100 mM MOPS with minimum salt was used. Online-monitored DO measurements were used to control the repeated additions of feeding solutions and both the feeding methods resulted in high cell densities of approximately 80 OD besides pH stabilization in between 6.5 and 7 for favorable growth of *E. coli*. Both the cultivations, one with high volume of glycerol and another with low volume of the same, exhibit high biomass concentration. Traditional bioreactors need huge amounts of energy for mixing, heating, and cooling during scale-up process but RoboLector microbioreactors are efficient and precise to hasten the development of bioprocesses using microbial cells [1].

2.2 Cultivation of *Pichia pastoris* using microbioreactors

Secreted proteins during bioprocesses undergo proteolysis and fragmentation; therefore, samples are generally removed from fermentation broth to obtain a kinetic growth curve. The purpose is to identify the optimum cultivation setup to reach maximum activity for the recombinant enzyme. RoboLectors are used to understand the kinetics of *Pichia pastoris* fed-batch cultivation [1]. Kinetics was monitored online by drawing 20 μl automatically and the concentrations of secreted enzymes were plotted against time to identify the space-time yield. During the automatic pipetting process, robotic tips immersed in the solution without any shaking, thus preventing artifacts of sedimentation.

To increase the productivity, softwares are programmed to provide outline of the experiments as per the input parameters. Dosing volume and variations in the concentrations of glycerol, ammonium hydroxide, and methanol can be given in the inputs. In 48 parallel fed-batch cultivations, a factorial design play gives a value, $2^4 + 3 = 19$; these different possibilities can be performed in one single analysis. Once the experiment is completed, the system automatically generates a contour plot and summarizes the activity profile of the recombinant enzymes with respect to the dosing volume and feeding solutions. For a particular recombinant product from *Pichia pastoris,* following factors were summarized at the completion of the experiment: 35% v/v methanol, 25% w/w ammonium hydroxide, and 150 g/l as feed composition for a dosing volume of 5 μl. The above-mentioned experiment is repeated for a particular design to identify and confirm the variable factors and increase the productivity [1].

2.3 Microbioreactors to produce monoclonal antibodies

Monoclonal antibodies are one of the major products among biotherapeutics. Organisms such as Chinese hamster ovary (CHO) and human embryonic kidney (HEK) cells are utilized to produce monoclonal antibodies, and several bioprocess parameters need to be optimized [12, 13] for their production. Composition of culture media, cell growth rate, and antibody growth rate need favorable physicochemical parameters such as pH, temperature, and dissolved oxygen. Physicochemical parameters are controlled to initiate and sustain recombinant expression in the CHO cell culture and to provide nutrients and relevant growth factors to bioreactors. During the production of monoclonal antibodies, possibilities of forming various variants of immunoglobulin (IgG) molecule and glycosylated IgG forms can be seen. Similar to monoclonal antibodies, other protein therapeutics also pass through issues such as oxygenation, amination, and degraded product molecules. Each of these above-mentioned processes is lengthy and consumes enormous time during research and development (R&D).

Microbioreactors, due to their miniature size and parallel testing, are needed to accelerate research and development (R&D) at a faster pace. Microbioreactors encompassed with sensors [14] that can hold volume in the range of 1–20 μl are ideal to monitor the bioprocess parameters. Microbioreactors need to be designed as per the needs of users. For example, for users to develop bioprocess procedures for mammalian cell cultures, the list of biological functions of mammalian cells and technical functions of bioreactors becomes essential.

Biological functions and requirements include: (a) nature of the cells (CHO or HEK cells) with concentration ranging from 10,000 to 10,000,000, (b) expression of proteins such as IgGs, (c) serum-free medium, and (d) culture time of 7–14 days [15].

Technical information parameters must include: (a) shaking for better mixing, (b) oxygen transfer value >100 h^{-1} to aerate the culture so that it can be extended to a large-scale process, (c) oxygen permeability <1%, (d) surface hydrophobicity (10°) along with a confocal microscope installed in situ without disturbing the reaction, (e) cell culture volume, (f) transportation of culture media to contained culture, and (g) measurement of chemical and physical conditions of culture and media [15].

Information function, namely online or offline information, is also essential: (a) online information about sensors to control and detect pH, temperature, and pO_2 is significant and (b) offline information such as monomers in culture media, forms of IgGs, excreted metabolites, products formed, and residual nutrients is needed for analysis of analytes. It is important that microbioreactors should be designed as per the specifications based upon the above requirements.

These specifications also cater to the requirements of microbioreactors in laboratories with 2–5-liter capacity. Therefore, such processes could be scaled up to laboratory level with effortless ease at reduced cost. CHO cell culture microbioreactors used to optimize the production of monoclonal antibodies are schematically shown in **Figure 2**. This is a part of Hubka-Eder map [16] focused on biological and technical functions. Hubka-Eder map shows the importance of integration among the subsystems, that is, the expression systems, cell line, medium with various technical and information functions for the design of the microbioreactors for the maximum production of monoclonal antibodies [15].

2.4 Microbioreactors in drug discovery

Organ-on-a-chip [17] is an ongoing research to hasten the process of drug discovery and development. In vitro drug screening and safety testing [18–20] are essential to minimize extensive studies on laboratory animals. Microbioreactors designed as per the need of the study will be highly beneficial to understand the potency and toxicity of the developed drug molecules. Drug molecules upon absorption get distributed to various organs such as lung, liver, gut, intestine, etc. These organs majorly consist of enzymes that can metabolize the drug molecules to metabolites [21] which can either be potent or toxic. After metabolism, the drug and its metabolites need to reach the target site for a particular time and get eliminated from the circulating system [22]. Kidneys play an important role in excreting drugs and its metabolites. Hence, developing various organs on a chip, programmed through a microbioreactor will be very helpful to understand the safety of drugs and metabolites.

Heart-on-a-chip is one such device which is developed to understand the efficacy of drug molecules with cardiac cells. Studies have proved that these devices have the capability to lend a helping hand in drug discovery and development process.

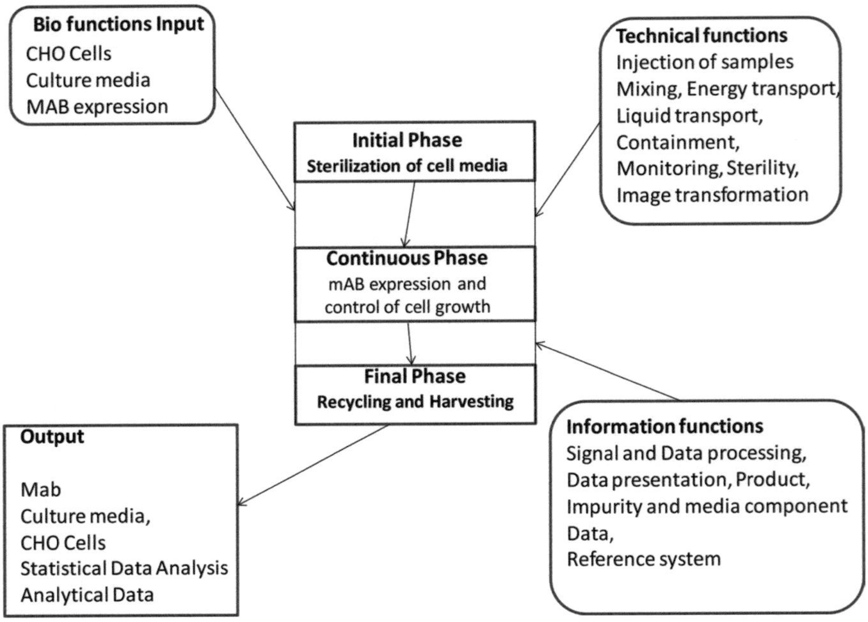

Figure 2.
Bioprocess optimization of a CHO cell culture to produce monoclonal antibodies.

3. Perfusion Bioreactors

3.1 Perfusion technology for biopharmaceutical production

The leading focus of R&D of any biopharma-based industry is to develop production process at reduced cost of production or rather work toward a minimal Cost of Goods (CoG) for cell culture-based products. Fed-batch mode of production has been the most tried and tested one, documented and dominantly followed methodology for production of biologics or biopharmaceuticals from mammalian cell culture. Due to the presence of expertise on this process within the industry, it has been proved to be the efficient process with judicious use of media coupled with a hassle-free downstream process. However, fed-batch mode has a severe limitation of inhibition of product formation and cell growth due to accumulation of inhibitory products during culture, especially ammonium ions, lactates, and proteases [23]. This leads to loss of key nutrients and incurs massive financial and resource loss to the production setup. This is especially significant if the product is a sensitive one. An alternative to this technology that came up in the early 1990s is perfusion technology. It involves addition of media or key media constituents at regular intervals along with retention of cells in the reactor and harvesting of formed products. However, in the initial phase, due to less advancement in media formulation and process development as well as dearth of efficient expression systems, not much advantage was offered by perfusion-based systems as compared to the well-established fed-batch systems and, hence, perfusion technology faced many failures and could not progress significantly. Many companies involved in manufacturing of perfusion-based systems especially for hybridoma-based monoclonal antibodies, like Endotronics, Cellex, and Biosyn, also folded up due to low market for perfusion bioreactors [24]. However, the pharma sector was on a constant lookout for newer or more effective production processes and hence advancements were continuously being made in perfusion technology. Now, with major advancements in media formulation, multiple technological options are available for cell retention and a rising trend has been witnessed in pharmaceutical industry for investment in adoption of new technologies and the cost effectiveness of perfusion-based processes, and their product yield increased significantly in comparison to conventional fed-batch processes [25]. Hence, this technology is on a comeback for the last few years and numerous life science companies are advocating the use of perfusion for biologicals production.

Factor VIII (ReFacto®) and IgG (Remicade® by Jansen Biotech and Simulect® by Novartis) are two of the leading products being produced commercially using perfusion technology [26, 27]. Apart from these, other products include interferon β-1a (Rebif from merck Sereno) and Ab Golimulab (Simponi® by Jansen Biotech), Factor VIII (Kogenate from Bayer Schering); anti-platelet MAb (ReoPro) and tumor necrosis factor MAb (Remicade from Centocor/J&J); CD52 a MAb (Campath from Genzyme/Sanofi); and a modified Factor VIII (Xyntha/ReFacto from Wyeth/Pfizer) among others. This substantiates the wide application of perfusion-based technologies for a variety of biologicals. The growing need of the pharma sector for reduced cost and enhanced productivity led to the resurgence of perfusion-based production technologies which have the key to revolutionize biopharmaceutical production process from mammalian cells.

3.2 Why perfusion technology?

Perfusion-based bioreactors are one of the upcoming reactor technologies based on continuous bioprocessing that offers the ease of continuous culturing of cells without nuisance of filter clogging or low throughput. In addition, there are less possibilities of waste accumulation and, hence, minimized chances of any product inhibition, especially while dealing with proteins prone to instability. Since nutrients are continuously exchanged and product harvest is maintained throughout along with cell retention, the availability of key media constituents is maintained consistently by providing host cells a stable environment leading to a high cell density and higher productivity with respect to desired compound. Typically, $3–10 \times 10^7$ cells/ml is the titer achieved using perfusion as compared to $5–25 \times 10^6$ in fed-batch cultures. Besides that, cost of goods in the 10,000–20,000-l fed-batch reactor is equivalent to that achieved using 1000 -l perfusion bioreactor [28]. Since the cells are subjected to a more stable and consistent environment, the recombinant proteins and other molecules produced are more like native compounds with similar glycosylation pattern and biological activity. This further increases the stability of the product and gives a high product yield. In a recent economical comparison between fed-batch and perfusion mode for the production of a glycosylated protein, CoG was analyzed and compared for both the processes based on BioSolve software. Continuous perfusion was calculated and found to be the most productive technology giving product at the rate of 265 kg/year as compared to 130 kg/year in fed-batch mode. Perfusion was also found to be the most cost-effective mode with the lowest overall CoG of $87/g as compared to $118/g for the continuous fed-batch process [29]. Also, there are less possibilities of failure and economic loss. Even if a problem is encountered, only the part being processed would need to be discarded, saving the rest for further processing. In a comparative study on production by both perfusion and fed-batch modes, CMC Biologics reported yield of 425 mg/l/day from perfusion bioreactor as compared to a yield of only 55 mg/l/day from fed-batch system for the same period of time [30]. Since most of the pharma companies thrive on economic profits, perfusion technology offers a lucrative mode of production especially as it beats the conventional fed-batch system in terms of productivity, efficiency, and capital investments.

3.3 Cell retention in perfusion

The prominent aspect which makes perfusion systems different and more valuable than fed-batch systems is the ability to yield a high cell mass due to the presence of cell retention devices. There are various ways through which cell retention is achieved [31]. Cells can be retained by making them grow inside bioreactor on hollow capillary fibers, flat plates, sponge-like materials, microcarrier particles, or other membranes. It can also be done by use of various cell separation devices like gravity-based cell settlers, spin filters, centrifuges, cross-flow filters, alternating tangential-flow filters, vortex-flow filters, acoustic settlers (sonoperfusion), and hydrocyclones [32]. Spin filter was one of the earliest available devices for cell retention which used a two-dimensional screen to retain the cells. However, it had limited scale-up potential especially in the scenario where rapid feed rate is needed. Gravity-based cell settlers are cost-effective but are marked by inefficient cell separation and significant cell loss, which lowers output and increases cost. Centrifuges have been known to give good performance but increase the production cost. Alternating tangential-flow filters (TFFs) have emerged as the most effective and practical means of high-density cell retention in a perfusion bioreactor [33]. The alternating tangential-flow action in these filters and location of diaphragm in the system prevent clogging as well as ensure a faster return of cells back to the reactor, bringing complete clarification. However, what need to be worked upon are other reactor specifications for handling large cell load at reduced volume and culture time. Also, scalability complications are a deterrent for many manufacturers. Many companies are targeting advancements in ATF system to handle increased cell load at smaller reactor volume. In a recent report on biologics development and manufacturing, the advancement in perfusion and its leading incorporation in manufacturing processes by leading biologics-based companies was attributed mostly to the advancements made in ATF systems which enhance the cell titers by multiple folds over extended periods of time, leading to higher volumetric productivity [34]. Acoustic wave separation (AWS) is another technology used by many companies for cell separation. Applikon Biotechnology and Pall Life Sciences are two such manufacturers advocating the use of acoustic waves to clump and settle down cells leading to their eventual separation. Sigma Aldrich Co. LLC (Merck & Co. Inc.), FiberCell Systems Inc., Zellwerk GmbH (Glen Mills in the United States), Cell Culture Company, ATMI Incorporated, PBS Biotech, Inc., GE Healthcare Life Sciences, Applikon Biotechnology, WAVE Life Sciences Biovest, AmProtein, Xcellerex, etc. are few of the leading manufacturers of perfusion bioreactors [35]. These reactors are revolutionizing the biopharmaceutical production industry and have established their presence in this sector preferably to stay for many years to come.

3.4 Perfusion and microbioreactors

Integration of perfusion technology with microbioreactors enhances the advantages associated with microbioreactors effectively. It further minimizes the losses associated with batch failure due to contamination. Even if contamination occurs earlier in the process, lesser media and other consumables would be wasted. However, the compatibility of the setup is amenable to technology development, scale-up, optimization, parameter sensitivity studies, and validation. Advancements in cell retention systems at microfluidic levels continue to be made. In a recent report, a novel microfluidic cell retention device based on inertial sorting was tested positively for retention of IgG1-producing Chinese hamster ovary (CHO) cell line. Parameters tested were cell retention efficiency, biocompatibility, and scalability. This was a spiral membrane-less system accomplishing cell retention based on hydrodynamic forces. The device was fabricated with polydimethylsiloxane (PDMS) and connected to spinner vessel-containing cells. There was also the flexibility of configuring the device to separate different-sized cells with a specific input flow rate. This gave an added advantage of flushing out non-viable cells, thus creating a healthier environment for the productive cell batch, leading to an increased productivity and product specificity [36]. In a first-of-its-kind study on feasibility of growth of human embryonic stem cells (hESCs), in continuous-flow microbioreactors, key cellular behavioral factors were assessed for robust cell growth under a range of flow rates. This would provide an assay platform for screening multiple cell lines for their capability to function in perfusion culture conditions as well as would aid in identifying optimum flow rates for their application in other microfluidic cell culture systems. Such microfluidic reactor setups would help in better understanding of stem cells maintenance and differentiation under varying stimulation conditions [37]. Perfusion microfluidic systems can also be used for growth and expression of proteins from bacterial cells. Growth of suspended cultures of *Pichia pastoris* in microbioreactor, integrated with perfusion devices, established the feasibility of use of such integrated systems for suspended bacterial growth and expression. Expression of recombinant human growth hormone (rhGH) or recombinant interferon alfa-2b (rIFNα-2b) by *Pichia pastoris* was found to be maintained after 11 days of growth which is close to the duration of many industrial processes with duration lasting from 7 to 30 days. Success of this study establishes the feasibility of use of perfusion-capable bioreactors to study impact of perfusion culture on expression and process optimization of many other microbial systems [38] at microfluidic levels, thus saving the initial process development costs [39].

3.5 Future of perfusion technology

Despite being in the sector for many years, perfusion technology did not gain much usage. However, with new methods for cell retention in the market and growing need for high throughput mammalian cell-based production systems, there is a renewed interest in the use of perfusion bioreactors. With a low capital and start-up cost, smaller setup, a smaller requirement for upstream and downstream processes, and reduced cost of failures, more and more compounds are being commercially produced using this technology. From leading biological manufacturers like Pfizer to upcoming ones like CMC Biologics, many companies are propagating the use of this technology with setup ranging from 250 to 2000 liters. Current research, however, would focus on scalability of the technology, creating a robust cell retention system, high-yield cell lines used in single-use technologies, cutting down on the initial establishment cost and creating the availability of expertise in perfusion-based devices and also the ease of bioprocessing modeling during continuous mode of operation. Also of relevance and on the rise is the use of single-use perfusion-based reactors for production. There are many manufacturers dealing with development of single-use bioreactors, sensors, and other systems to simplify setup and operation of perfusion cell culture. These include among Pall Life Sciences, Applikon Biotechnology, Sartorius, AcuSyst, GE Healthcare, and many others.

4. Conclusion

Microbioreactors are essential for high throughput screening of various strains and optimize the bioprocess development in industries. Microbioreactors assist in identifying the appropriate experimental conditions to scale up the production process. Production of biopharmaceuticals, enzymes, proteins, and medicinally important chemical compounds from organisms was attempted successfully. Microbioreactors have been demonstrated as a suitable technology to cultivate *Pichia pastoris*, *Escherichia coli*, monoclonal antibodies, etc., at large scale that are important for biopharmaceutical industries. Incorporation of microbioreactors in drug discovery and development will reduce the cost and time for developing therapeutic molecules. The growing need for a high production process and efficient production strategy has also championed the cause of perfusion-based technology interfaced with microbioreactors. Perfusion microfluidic bioreactors have been extremely beneficial in maintaining stem cells, expression of proteins, growth of cells, and microbial cultures. Thus, microbioreactors aid in developing high-quality products at affordable cost with minimal resources.

Acknowledgements

Authors thank Symbiosis International (Deemed University) for funding and research facility.

Conflict of interest

None declared.

Author details

Selvan Ravindran[1*], Pooja Singh[1], Sanjay Nene[2], Vinay Rale[1], Nutan Mhetras[1] and Anuradha Vaidya[1]

1 Symbiosis School of Biological Sciences, Symbiosis International (Deemed) University, Pune, India

2 Innovation Biologicals Private Limited, Pune, India

*Address all correspondence to: selvan_ravindran@yahoo.com

References

[1] Kensy F, Hemmerich J. Automation of microbioreactors—operating 48 parallel fed-batch fermentations at microscale. Bioprocess International. 2013;**11**(8):68-76

[2] Adrio JL, Demain AL. Microbial enzymes: Tools for Biotechnological processes. Biomolecules. 2014;**4**(1):117-139

[3] Hemmerich J, Kensy F. Automated microbioreactor systems for pharmaceutical bioprocessing: Profiling of seeding and induction conditions in high throughput fermentations. Pharmaceutical Bioprocessing. 2014;**2**(3):227-235

[4] Erickson B, Winters PN. Perspective on opportunities in industrial biotechnology in renewable chemicals. Biotechnology Journal. 2012;**7**(2):176-185

[5] Unthan S, Radek A, Wiechert W, Oldiges M, Noack S. Bioprocess automation on a mini pilot plant enables fast quantitative microbial phenotyping. Microbial Cell Factories. 2015;**14**:1-11

[6] Noack S, Wiechert W. Quantitative metabolomics: A platform? Trends in Biotechnology. 2014;**32**(5):238-244

[7] Hemmerich J, Noack S, Weichert W, Oldiges M. Microbioreactor systems for accelerated bioprocess development. Biotechnology Journal. 2018;**13**:e1700141

[8] Toeroek C, Puschmann MC, Bayer K, Striedner G. Fed-batch like cultivation in a micro-bioreactor: Screening conditions relevant for *Escherichia coli* based production processes. SpringerPlus. 2015;**4**(490):1-10

[9] 32 parallel microbioreactors pH control continuous feeding online monitoring scalability, Application note, 1-6. Available from: www.m2p-labs.com

[10] Muhlman M, Kunze M, Ribeiro J, Geinitz B, Lehmann C, Schwaneberg U, et al. Cellulolytic robolector-towards an automated high-throughput screening platform for recombinant cellulose expression. Journal of Biological Engineering. 2017;**11**(1):1-18

[11] Betts JPJ, Warr SRC, Finka GB, Uden M, Town M, Janda JM, et al. Impact of aeration strategies on fed-batch cell culture kinetics in a single-use 24-well miniature bioreactor. Biochemical Engineering Journal. 2014;**82**:105-116

[12] Janakiraman V, Kwiatkowski C, Kshirsagar R, Ryll T, Huang YM. Application of high-throughput mini-bioreactor system for systematic scale-down modeling, process characterization and control strategy development. Biotechnology Progress. 2015;**31**:623-632

[13] Huang YM, Hu WW, Rustandi E, Chang K, Makagianser YK, Ryll T. Maximizing productivity of CHO cell based fed-batch culture using chemically defined media conditions and typical manufacturing equipment. Biotechnology Progress. 2010;**26**:1400-1410

[14] Kommenhoek EE, Van Leeuwen M, Gardeniers H, Van Gulik WM, Van den Berg A, Li X, et al. Lab-scale fermentation tests of microchip with integrated electrochemical sensors for pH, temperature, dissolved oxygen and viable biomass concentration. Biotechnology and Bioengineering. 2008;**99**:884-892

[15] Mandenius CF. Conceptual design of micro-bioreactors and organ-on-chips for studies of cell cultures. Bioengineering. 2018;**5**(56):1-22

[16] Derelov M, Detterfelt J, Bjorkman M, Mandenius CF. Engineering design methodology for bio-mechatronic products. Biotechnology Progress. 2008;**24**:232-244

[17] Marsano A, Conficconi C, Lemme M, Occhetta P, Gaudiello E, Votta E, et al. Beating heart on a chip: A novel microfluidic platform to generate functional 3D cardiac microtissues. Lab on a Chip. 2016;**16**:599-610

[18] Ravindran S, Rokade R, Suthar J, Singh P, Deshpande P, Khambadkar R, et al. *In vitro* biotransformation in drug discovery. In: Boparalla V, editor. Drug Discovery—Concepts to Market. London, United Kingdom: InTech Open; 2018. pp. 1-13

[19] Ravindran S, Jadhav A, Surve P, Lonsane G, Honrao P, Nanda B. Technologies and strategies to characterize and quantitate metabolites in drug discovery and development. Biomedical Chromatography. 2014;**28**:1547-1553

[20] Surve P, Ravindran S, Acharjee A, Rastogi H, Basu S, Honrao P. Metabolite characterization of anti-cancer agent Gefitinib in human hepatocytes. Drug Metabolism Letters. 2013;**7**:126-136

[21] Ravindran S, Honrao C, Sahu R, Basit A, Madireddy S, Basu S, et al. Optimal use of mass spec scan modes to identify an unknown metabolite. Drug Discovery Today. 2011;**3**:259-261

[22] Ravindran S, Suthar JK, Rokade R, Pratinidhi A, Kambadkar R, Utekar S. Pharmacokinetics, metabolism, distribution and permeability of nanomedicine. Current Drug Metabolism. 2018;**19**(4):327-334

[23] Bielser JM, Wolf M, Souquet J, Broly H, Morbidelli M. Perfusion mammalian cell culture for recombinant protein manufacturing–a critical review. Biotechnology Advances. 2018;**36**(4):1328-1340

[24] Langer ES. Trends in perfusion bioreactors. Bioprocess International. 2011;**9**(4):18-22

[25] Langer ES, Rader RA. Continuous bioprocessing and perfusion: Wider adoption coming as bioprocessing matures. Bioprocessing Journal. 2014;**13**:43-49

[26] Farid SS. Established bioprocesses for producing antibodies as a basis for future planning. In: Scheper T, editor. Advances in Biochemical Engineering/ Biotechnology. Vol. 2006. Heidelberg: Springer. pp. 1-42

[27] Kelley B, Jankowski M, Booth J. An improved manufacturing process for Xyntha/ReFacto AF. Haemophilia. 2010;**16**(5):717-725

[28] Compton BJ. Use of Perfusion Technology on the Rise. 2007. Available from: https://www.genengnews.com/magazine/use-of-perfusion-technology-on-the-rise/ [Accessed: 21-11-2018]

[29] Lim J, Sinclair A, Shevitz J, Bonham-Carter J. An economic comparison of three cell culture techniques. BioPharm International. 2011;**24**(2):54-60

[30] Carstens JN, Clarke HRG and Jensen JP. Perfusion! Jeopardy or the Ultimate Advantage? CMC Biologics. September, 2009. Available from: http://www.cmcbio.com/ Portals/0/CMC/docs/perfusion.pdf [Accessed: 21-11-2018]

[31] Voisard D, Meuwly F, Ruffieux PA, Baer G, Kadouri A. Potential of cell retention techniques for large-scale high-density perfusion culture of suspended mammalian cells. Biotechnology and Bioengineering. 2003;**82**(7):751-765

[32] Woodside SM, Bowen BD, Piret JM. Mammalian cell retention devices for stirred perfusion bioreactors. Cytotechnology. 1998;**28**(1-3):163-175

[33] Carter JB and Shevitz J. A Brief History of Perfusion Biomanufacturing. 2011. Available from: https://pdfs.semanticscholar.org/0131/5d5a12ed010e8ee20882cfa8dfbf6546484f.pdf [Accessed: 21-11-2018]

[34] Mahler G. Advancing Biologics development and Manufacturing. 2016. Available from: http://www.agcbio.com/resource-center/news/advancing-biologics-development-and-manufacturing [Accessed: 26-11-2018]

[35] Perfusion Bioreactors Market: Global Industry Analysis and Forecast 2017-2025. Available from: https://www.persistencemarketresearch.com/market-research/perfusion-bioreactors-market.asp [Assessed: 27-11-2018]

[36] Kwon T, Prentice H, De Oliveira J, Madziva N, Warkiani ME, Hamel JFP, et al. Microfluidic cell retention device for perfusion of mammalian suspension culture. Scientific Reports. 2017;7(1):6703. DOI: 10.1038/s41598-017-06949-8

[37] Titmarsh D, Hidalgo A, Turner J, Wolvetang E, Cooper-White J. Optimization of flowrate for expansion of human embryonic stem cells in perfusion microbioreactors. Biotechnology and Bioengineering. 2011;**108**(12):2894-2904

[38] Rokade R, Ravindran S, Singh p SJ. Microbial biotransformation for the production of steroid medicament. In: Vijayakumar R, Raja SSS, editors. Secondary Metabolites—Sources and Applications. London, United Kingdom: InTech Open; 2018. pp. 115-124

[39] Mozdzierz NJ, Love KR, Lee KS, Lee HL, Shah KA, Ram RJ, et al. A perfusion-capable microfluidic bioreactor for assessing microbial heterologous protein production. Lab on a Chip. 2015;**15**(14):2918-2922

Chapter 8

Integration of Membranes and Bioreactors

Katalin Belafi-Bako and Peter Bakonyi

Abstract

Combined application of bioreactors and membrane separations are considered as membrane bioreactors (MBRs). Examples for the application of MBRs are given in this chapter both for large scale (wastewater treatments) and in other areas in smaller scale. Wastewater treatments are the majority of the large-scale applications, where biological degradation is coupled with membrane filtration (microfiltration and ultrafiltration). Other types of MBRs include integration of biotransformations and bioconversions by microorganisms and enzymes with membrane separation processes, not only with filtration but also with pervaporation, electrodialysis, and gas separation. These MBRs provide significant advantages compared to the conventional batch bioprocesses. In this chapter, several examples are presented for both applications.

Keywords: ultrafiltration, pervaporation, electrodialysis, gas separation, biocatalysis

1. Introduction

"As per definition, the bioreactor is the designed space where biochemical reactions take place" [1]. If some compounds should be removed from the bioreactor, a separation step can be connected to the reactor. Among separation processes, membrane techniques are especially attractive since they operate under mild conditions, they are easy to combine and vary, the scale-up is simple due to the modular construction, and they do not need and produce hazardous materials (environmental-friendly processes) [2–4].

Application of membranes integrated in bioreactions is often considered as membrane bioreactors (MBRs). In the literature [5–8] the term *MBRs* itself is referred almost exclusively for various wastewater treatments. Concerning full-scale applications, it seems correct. In these technologies only pressure-driven membrane processes such as microfiltration (MF), ultrafiltration (MF), nanofiltration (NF), and reverse osmosis (RO) have been used, although there are numerous other membrane separation techniques that are available nowadays, which can be inserted into bioprocesses.

In this work the first section is going to summarize the "conventional" applications of MBRs, i.e., wastewater treatments, while in the second part, some other types of MBRs will be presented where—beyond pressure-driven processes—some more membrane separation methods, like pervaporation (PV), gas separation (GS), electrodialysis (ED), etc., will be presented.

2. Large-scale applications

The usage of membrane bioreactors means a well-established technology for several types of wastewater treatments. The recently worldwide opened (of will opened soon) large-scale—over 100,000 m^3/d capacity—MBR plants for municipal wastewater treatment are summarized in [8]. Other main application areas are as follows [6, 8]:

- Industrial wastewater treatment
- Landfill leachate treatment
- Sludge digestion
- Treatment of human excrement

In these applications MBRs are considered as complex systems integrating biological degradation of waste products with membrane filtration [6]. Thus an MBR is composed of two parts: a biological unit and a membrane module. According to the location of the membrane module (from the aspect of architecture), the MBRs can be classified into two groups: internal and external systems. The internal system (frequently called submerged MBR) of the membrane module is placed in the bioreactor [6]. Outer skin membranes are applied, and the permeate side is under suction (vacuum); moreover, aeration and mixing can be easily achieved, as well. In the external MBR, the membrane is located outside the bioreactor, and the treated wastewater is recirculated through the primary side of the membrane module. The driving force is provided by the pressure from the high cross flow volumetric rate along the membrane surface [6].

Regarding configuration there have been numerous types of modules (plate and frame, tubular, rotary disk, hollow fiber) and membranes (cellulose acetate, polyethylene, polysulfone, polyolefin, metallic, ceramic—mainly in ultrafiltration and microfiltration range) applied.

MBRs are widely considered as an effective technology in removing both organic and inorganic contaminants in wastewater, have a good control of biological activity—the effluent is generally free of bacteria and pathogens—need smaller plant size, and provide higher organic loading rates [5, 6]. Beyond the benefits, however, there are some serious drawbacks and issues which should be enhanced, like membrane fouling and energy consumption. To solve these problems and to widen the application opportunities, several special techniques and methods have been investigated and suggested, where unusual environments (e.g., varying the level of oxygen, flow pattern, airlift) and unique procedures (e.g., applying electric power, magnetic effects, illumination, vibration, osmosis) are added and/or applied for the MBR systems. Some of them are listed in **Table 1**, together with their abbreviations.

Anaerobic degradation of organic wastes [10, 11]—similar to the classical aerobic wastewater treatments—can be carried out by microbial consortia. The technology results in gas products ("biogas"). It consists usually of methane, CO2, and H2S mainly. To separate these compounds, membranes can be applied, as well, but these membranes have selectivity toward one of the compounds in the gaseous mixture [12, 15]. The process is called membrane gas separation, and its driving force is mainly the pressure difference, similar to UF and MF.

Recently biogas plants are built in connection with wastewater plants to complete the degradation process and to obtain energy which may cover the energy demand of the process; moreover other types of wastes (slaughter wastes,

Name	Abbr.	Ref
Anaerobic-anoxic-oxic membrane bioreactor	A^2O MBR	[8]
Anaerobic fluidized bed membrane bioreactor	AFMBR	[9]
Anaerobic membrane bioreactor	AnMBR	[6–11]
Airlift oxidation ditch membrane bioreactor	AOXMBR	[8]
Bioelectrochemical membrane reactor	BEMR	[8]
Batch granulation membrane bioreactor	BG-MBR	[8]
Baffled membrane bioreactor	BMBR	[10]
Electrochemical membrane bioreactor	EMBR	[8]
Gas separation—membrane bioreactor	GS-MBR	[12]
Hybrid-growth membrane bioreactor	HG-MBR	[9]
Immersed hollow fiber membrane bioreactor	IHFMB	[13]
Membrane electro-bioreactor	MEBR	[8]
Membrane gradostat reactor	MGR	[14]
Magnetically induced membrane vibration membrane bioreactor	MMV-MBR	[8]
Membrane photobioreactor	MPBR	[8]
Osmotic membrane bioreactor	OMBR	[7]
Reciprocation membrane bioreactor	rMBR	[8]
Single fiber membrane gradostat reactor	SFMGR	[14]

Table 1.
Various unusual membrane bioreactors.

agricultural wastes) can be degraded, as well. In these complex plants, MBRs can be applied in both aerobic (coupled with pressure-driven membrane techniques) and anaerobic (coupled with membrane gas separation) systems.

3. Novel applications in developing stage

3.1 Types of MBRs and classifications

Integrated systems where membranes are combined with the bioreactions—other than wastewater degradation—can be classified similarly: external and integral setups. However, it is difficult to connect certain types of membrane processes (e.g., pervaporation, electrodialysis) to the bioreactor *externally*; thus they are used as *internal* systems. **Figure 1** presents an example for internal MBR, where the configuration is illustrated. These systems have the advantage to handle (e.g., disconnect easily) the membrane module independently from the bioreactor.

On the other hand, there are successful examples for external (immersed membranes) MBR systems, which can be applied in special cases, e.g., for manufacture of pharmaceutical intermediaries. Loh and colleagues reported [13] that the biotransformation of indene to cis-indandiol was achieved in an IHFMB system resulting in higher effectiveness.

MBR systems can be distinguished according to the membrane process integrated [16, 17]. Beyond pressure-driven methods (microfiltration, ultrafiltration, nanofiltration), other techniques like pervaporation, electrodialysis, and gas

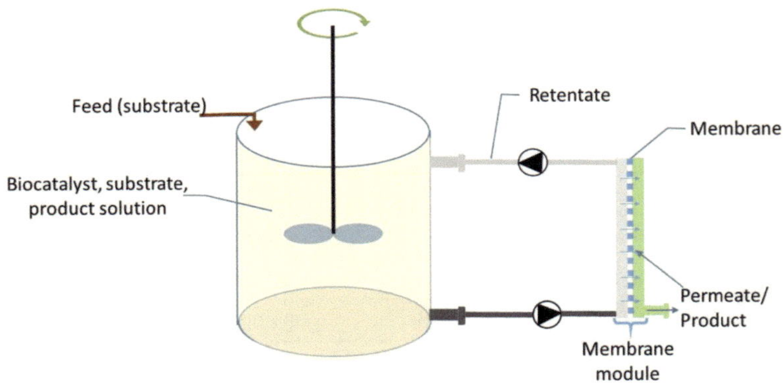

Figure 1.
Outline of an internal (integrated) system.

separation can be applied. To connect these membrane processes to the bioreaction, careful design and optimization should be accomplished before starting the operation of MBRs. Plate and frame, tubular, as well as hollow fiber modules can be used in the MBRs.

The novel MBR systems can be further categorized regarding the biocatalyst used: enzymes as well as microbes (moreover microbial consortia) can be applied. The biocatalysts can be further classified concerning the state of the biocatalysts, as well (free or immobilized); moreover they can be immobilized onto the membrane (which is considered as *catalytic active membrane* [18]) or on other supports. Some examples are presented here for all the classes.

3.2 Enzymatic MBRs

Enzymatic MBRs have been used mainly in hydrolytic reactions where the substrates are, e.g., triacylglycerols and polysaccharides (starch, cellulose, pectin), but some other reactions occur, as well, like esterification. **Table 2** summarizes the important features of these systems.

Hydrolysis of triacylglycerols (fats and oils) results in glycerol and fatty acids (long chain carboxylic acids). The higher demand of fatty acids in various industrial sectors (e.g., production of cosmetics, detergents) made the hydrolysis an important process recently. The main difficulty of the process is that the two reactants, triacylglycerols and water, are not miscible. When the reaction is carried out by enzymatic catalysis [17], numerous advantages are provided: it takes place

Reaction	Substrate	Enzyme	State	Membrane	Ref.
Hydrolysis	Triacylglycerol	Lipase	Immobilized	UF	[19–21]
Hydrolysis	Protein	Protease	Free and immobilized	UF	[18]
Hydrolysis	Starch	Amylases	Free	UF	[22, 23]
Hydrolysis	Cellulose	Cellulases	Immobilized	UF	[24]
Hydrolysis	Pectin	Pectinases	Free	UF ED	[25–27] [28–31]
Esterification	Acids and alcohols	Lipases	Immobilized	PV	[32–35]

Table 2.
Examples for enzymatic MBRs.

under mild conditions, instead of high temperature and pressure, implying energy saving and better quality products. Accomplishing the hydrolysis in a MBR, further benefits are added to the process: the two phases remain separated during the reaction and there is no emulsion formed and no separation needed, since the products are separated by the membrane.

The enzyme suitable for the hydrolysis is called lipase. It can be immobilized onto the membrane. During the reaction, the pure triacylglycerol substrate (no solvent is present) and water are flown separately into the two sides of the membrane. Both hydrophilic and hydrophobic membranes can be applied [19]. In case of hydrophilic membrane, the substrate should be circulated on the enzyme side of the membrane, while water is on the other side. Thus fatty acids formed are remaining in the oily phase (and accumulated there), while glycerol is passing through the hydrophilic membrane into the water phase. Applying hydrophobic membrane the arrangement is the other way around: the substrate is flown on the nonenzyme side of the membrane. In our experiments lipase from *Candida rugosa* (Sigma) and hydrophilic cellulose acetate membrane (cut off 3000) was used [19] successfully for the hydrolysis of various oils and fats [20, 21].

In hydrolytic degradation of polymers (polysaccharides and proteins), the small-sized products (e.g., glucose) often have inhibitory effect, and thus their removal is beneficial for the reaction. MBRs provide a simple solution, since the membrane applied serves for rejection of the long polymer chains as well as the biocatalysts (enzymes or cells), while the product can easily pass through the membrane.

Hydrolysis of various proteins—e.g., from milk, whey, plasma—is an important step in industrial processes, and it is realized by protease enzymes. The MBRs provide significant advantages for the process [18] including retention and reuse of the biocatalysts, avoiding product inhibition—a possibility for continuous operation.

Regarding enzymatic hydrolysis of polysaccharides, starch, cellulose, and pectin are considered. Hydrolytic products of starch are utilized widely in food industry (e.g., maltodextrin, dextrose, and high fructose syrups). The reaction is catalyzed by amylase enzymes. When the hydrolysis is carried out in MBRs, higher effectivity can be achieved [22, 23]. In our work hollow fiber cellulose acetate (UF) membrane module (jacketed) was used, and the experimental results confirmed that not only purified amylase preparations but a crude fermentation filtrate, an enzyme complex solution (mainly glucoamylase) produced by *Aspergillus awamori*, was capable for degradation of starch effectively and continuously.

Cellulose is a long-chain polysaccharide containing glucose (monomer)—similarly to starch. Enzymatic hydrolysis of cellulose, however, is more difficult [24]. The reaction can be carried out in MBRs to overcome some of the problems. A special tubular membrane module was used to carry out the reaction, where a fine, hairy surface membrane was built in [24]. The hydrolysis was catalyzed by a Celluclast preparation (Novozymes), and it was found that the special MBR had a positive effect on the process and enhanced conversion and productivity of the enzyme reaction was accomplished.

Pectin is a polysaccharide occurring in the cell walls of certain plants, mainly in the fruits like apple, orange, some berried, etc. It is important to use pectolytic enzymes in processing of fruits (production of fruit juices) [25, 26] to enhance yields and improve liquefaction and clarification. As a result of the hydrolytic degradation of pectin, galacturonic acid (monomer) is formed. It was assumed that the hydrolysis was inhibited by the monomer; therefore MBRs seemed a promising reactor type to realize the process effectively. Regenerated cellulose UF membrane (30 kD) and polygalacturonase enzyme preparation (Sigma) from *Aspergillus niger* were used in the experiments, and it was found that the MBR worked reliably with excellent stability for long and higher volumetric productivity was achieved than the batch system. To enhance the effectiveness of the process, vacuum was applied in the secondary side of the membrane [27].

The reaction between acids and alcohol results in esters and water. It can be catalyzed by lipase enzymes. The reaction is reversible and product removal enhances the yield. Both products can be separated from the reaction mixture by membranes. If the substrates are low molecular weight acids and alcohols, flavor esters are formed, which are volatile compounds. Thus pervaporation (PV) can be applied for their recovery that is quite easy to connect online to the reaction, forming an integrated MBR system. Ethyl acetate [32] and other flavor esters [33] were manufactured effectively by a lipase preparation (*Candida antarctica* immobilized onto a resin, from Novozymes) and separated online by PV using hydrophobic membranes, and the permeate was obtained by using vacuum traps cooled by dry ice.

PV is not suitable for the separation of higher molecular weight ester products, but it is possible to apply it for removal of water formed during the esterification. Ester type biolubricant from oleic acid and alcohols [34] and ester of 2-choloropropionic acid and 1-butanol [35] were manufactured by using integrated PV for water removal.

Electrodialysis also can be combined to enzymatic processes, where charged compounds are formed and they can be separated online. This kind of MBR was applied for continuous recovery of galacturonic acid obtained from the hydrolysis of pectin by a pectinase preparation [28]. Both monopolar and bipolar membranes were successfully used [29, 30].

Summarizing the application possibilities of MBRs in enzymatic processes, it can be concluded that they provided real advantages for the bioprocesses listed above and—beyond process intensification [31, 36]—these systems can be operated in continuous mode of operation: constant uptake of substrate (and biocatalyst, if necessary) and release of product can be achieved.

3.3 Microbial MBRs

In microbial MBRs a single microorganisms can work, but sterile conditions should be provided. One of the examples given here is the biohydrogen production, where gas separation is connected and the other one is manufacture of organic acids, where electrodialysis is used for in-situ separation of the products.

Biohydrogen can be produced by fermentation using various microbes. Some of them need light for the operation (photofermentation), while others do not (dark fermentation). *Escherichia coli* belong to the dark fermentative group and is able to form biogas containing mainly hydrogen and CO2 in the headspace of the bioreactor. When the bioreactor is integrated by a membrane gas separation unit (MBR), higher productivity was possible to achieve, as proven in Veszprem [37–42]. Polyimide hollow fiber membranes and *E. coli* XL-1 Blue strain were used for the experiments, which resulted in a feasible concept for the integrated production and purification of the promising energy carrier, hydrogen gas [43–46]. Additionally, hydrogen fermenters could be attached to microbial electrochemical technologies, thus giving opportunity for adequate treatment of the effluents containing residual organic matter [47–50].

Transformation of fumaric acid into L-malic acid—which is the second most popular food acid—is an equilibrium reaction; thus, product removal could increase the yield. The reaction is catalyzed by immobilized cells of *Leuconostoc* and *Brevibacterium* species containing fumarase enzyme, and the acid (in salt form) can be separated by electrodialysis [51] integrated to the reaction.

Malic acid is not the only acid that can be produced in integrated (MBR) system, other organic acids—manufactured by fermentation—can be online separated by monopolar or bipolar electrodialysis systems.

A relatively novel area of MBR applications is the immobilization and cultivating of microbes; the system is called membrane gradostat reactor (MGR) [15]. The biofilm growing (and attached) on the surface has a significant effect on the permeability. The behavior of these systems was studied in case of, e.g., *Streptomyces coelicolor*, in a vertically oriented capillary MGR.

4. Conclusions

Bioprocesses integrated with membrane separations (MBRs) are able to provide more effective, successful bioconversions and bioreactions. The MBRs have been used in large scale worldwide in wastewater treatments, on one hand, while there are other, special applications in smaller scale (and sometimes only in a developing stage), on the other hand. Examples for both areas were given in this chapter.

Acknowledgements

The research work was partly supported by the project EFOP-3.6.1-16-2016-00015. Peter Bakonyi acknowledges the support received from National Research, Development and Innovation Office (Hungary) under grant number PD 115640.

Abbreviations

ED	Electrodialysis
GS	Gas separation
MBRs	Membrane bioreactors
MF	Microfiltration
NF	Nanofiltration
PV	Pervaporation
RO	Reverse osmosis
UF	Ultrafiltration

Author details

Katalin Belafi-Bako* and Peter Bakonyi
University of Pannonia, Veszprem, Hungary

*Address all correspondence to: bako@almos.uni-pannon.hu

References

[1] Mandenius CF, editor. Bioreactors: Design, Operations and Novel Applications. 1st ed. Oxford, UK: Verlag-Wiley; 2016. DOI: 10.1002/9783527683369

[2] Drioli E, Giorno L. Comprehensive Membrane Science and Engineering (I-IV). 1st ed. Amsterdam, NL: Elsevier; 2010

[3] Baker RW. Membrane Technology and Applications. Wiley; 2012

[4] Purkait MK, Singh R. Membrane Technology in Separation Science. Boca Raton: CRC Press; 2017

[5] Yoon SH. Membrane Bioreactor Processes. Boca Raton: CRC Press; 2016

[6] Cicek N. A review on membrane bioreactors and their potential application in the treatment of agricultural wastewater. Canadian Biosystems Engineering. 2003;**45**:637-649

[7] Huang L, Lee DJ. Membrane bioreactor: A mini review on recent R+D works. Bioresource Technology. 2015;**194**:383-388. DOI: 10.1016/j.biortech.2015.07.013

[8] Krzeminski P, Leverette L, Malamis S, Katsou E. Membrane bioreactors – A review on recent developments in energy reduction, fouling control, novel configurations, LCA and market prospects. Journal of Membrane Science. 2017;**527**:207-227. DOI: 10.1016/j.memsci.2016.12.010

[9] Shin C, McCarty PL, Kim J, Bae J. Pilot scale temperature-climate treatment of domestic wastewater with a staged anaerobic fluidized membrane bioreactor. Bioresource Technology. 2014:95-103. DOI: 10.1016/j.biortech.2014.02.060

[10] Dvorak L, Gomez M, Dolina J, Cernin A. Anaerobic membrane bioreactors – A mini review with emphasis on industrial wastewater treatment: Applications, limitations and perspectives. Desalination and Water Treatment. 2016;**57**:19062-19076. DOI: 10.1080/19443994.2015.1100879

[11] Szentgyörgyi E, Bélafi-Bakó K. Anaerobic membrane bioreactors. Hungarian Journal of Industrial Chemistry. 2010;**38**:181-185

[12] Bakonyi P, Kumar G, Bélafi-Bakó K, Sang-Hyoun Kim SH, Koter S, Kujawski W, et al. A review of the innovative gas separation membrane bioreactor with mechanisms for integrated production and purification of biohydrogen. Bioresource Technology. 2018;**270**:643-655. DOI: 10.1016/j.biortech.2018.09.020

[13] Cheng XY, Tong YW, Loh KC. An immersed hollow fiber membrane bioreactor for enhanced biotransformation of indene to cis-indandiol using pseudomonas putida. Biochemical Engineering Journal. 2014;**87**:1-7. DOI: 10.1016/j.bej.2014.03.011

[14] Godongwana B, De Jager D, Sheldon MS, Edwards W. The effect of Streptomyces ceolicolor development on the hydrodynamics of a vertically oriented capillary membrane gradostat reactor. Journal of Membrane Science. 2009;**333**:79-87. DOI: 10.1016/j.memsci.2009.01.051

[15] Szentgyörgyi E, Nemestóthy N, Bélafi-Bakó K. Application of membranes in biogas production. Desalination and Water Treatment;**210**(14):112-115. DOI: 10.5004/dwt.2010.2582

[16] Bélafi-Bakó K, Gubicza L. Biocatalysts and membranes. In: Bélafi-Bakó K, Gubicza L, Mulder M, editors. Integration of Membrane

Processes into Bioconversions. London: Kluwer Academic; 2000. pp. 131-140

[17] Bélafi-Bakó K. Simultaneous application of enzymes and membranes in food processing. In: Jerrod MC, editor. Food Engineering Research Trends. New York: Nova Science Publishers; 2008. pp. 263-279

[18] Rios GM, Belleville MP, Paolucci D, Sanchez J. Progress in enzymatic membrane reactors—Review. Journal of Membrane Science. 2004;**242**:189-196. DOI: 10.1016/j.memsci.2003.06.004

[19] Bélafi-Bakó K, Dombi Á, Szabó L, Nagy E. Triacylglycerol hydrolysis by lipase in a flat membrane bioreactor. Biotechnology Techniques. 1994;**8**:671-674

[20] Csányi E, Bélafi-Bakó K. Semi-continuous fatty acid production by lipase. Hungarian Journal of Industrial Chemistry. 1999;**27**:293-295

[21] Keöves E, Csányi E, Bélafi-Bakó K, Gubicza L. Membrane reactors for biochemical reactions. Hungarian Journal of Industry and Chemistry. 1999:41-42

[22] Koutinas A, Bélafi-Bakó K, Kabiri-Badr A, Tóth A, Gubicza L, Webb C. Enzymatic hydrolysis of polysaccharides, hydrolysis of starch by an enzyme complex from fermentation by *Aspergillus awamori*. Food and Bioproducts Processing. 2001;**79**:41-45. DOI: 10.1205/09603080151123353

[23] Bélafi-Bakó K, Nemestóthy N, Milisic V, Gubicza L. Membrane bioreactor for utilisation of carbohydrates in waste streams. Desalination. 2002;**149**:329-330. DOI: 10.1016/S0011-9164(02)00803-2

[24] Bélafi-Bakó K, Koutinas A, Nemestóthy N, Gubicza L, Webb C. Continuous enzymatic cellulose hydrolysis in a tubular membrane bioreactor. Enzyme and Microbial Technology. 2006;**38**:155-161. DOI: 10.1016/j.enzmictec.2005.05.012

[25] Bélafi-Bakó K, Eszterle M, Kiss K, Nemestóthy N, Kovács S, Gubicza L. Utilisation of a membrane bioreactor for pectin hydrolysis by Aspergillus Niger polygalacturonase. Desalination. 2006;**200**:507-508. DOI: 10.1016/j.desal.2006.03.414

[26] Bélafi-Bakó K, Eszterle M, Kiss K, Nemestóthy N, Gubicza L. Hydrolysis of pectin by Aspergillus Niger polygalacturonase in a membrane bioreactor. Journal of Food Engineering. 2007;**78**:438-442. DOI: 10.1016/j.jfoodeng.2005.10.012

[27] Kiss K, Nemestóthy N, Gubicza L, Bélafi-Bakó K. Vacuum assisted membrane bioreactor for enzymatic hydrolysis of pectin from various sources. Desalination. 2009;**241**:29-33. DOI: 10.1016/j.desal.2008.01.055

[28] Molnár E, Eszterle M, Kiss K, Nemestóthy N, Fekete J, Bélafi-Bakó K. Utilisation of electrodialysis for galacturonic acid recovery. Desalination. 2009;**241**:81-85. DOI: 10.1016/j.desal.2008.01.059

[29] Molnár E, Nemestóthy N, Bélafi-Bakó K. Galacturonic acid recovery from pectin rich agro-wastes by electrodialysis with bipolar membranes. Hungarian Journal of Industrial Chemistry. 2008;**36**:95-99. DOI: 10.1515/186

[30] Belafi-Bako K, Molnar E, Csanadi Z, Nemestothy N. Comparative study on electrodialysis systems for galacturonic acid recovery. Hungarian Journal of Industrial Chemistry. 2012;**40**:65-67. DOI: 10.1515/343

[31] Satyawali Y, Vanbroekhoven K, Dejonghe W. Process intensification: The future of enzymatic processes? Biochemical Engineering Journal. 2017;**121**. DOI: 10.1016/j.bej.2017.01.016

[32] Bélafi-Bakó K, Kabiri-Badr A, Szamosszegi V, Keöves E, Koszorz Z, Gubicza L. Production of natural flavour esters by lipase in integrated system. Mededelingen van de Faculteit Landbouwwetenschappen, Universiteit Gent. 1999;**64/5a**:311-314

[33] Bélafi-Bakó K, Kabiri Badr A, Nemestóthy N, Ehrenstein U, Gubicza L. Kinetics of ethyl acetate production by lipase in organic solvent and solvent-free system. Chemical Papers. 2003;**57**:277-280

[34] Bélafi-Bakó K, Dörmő N, Ulbert O, Gubicza L. Application of pervaporation for removal of water produced during enzymatic esterification in ionic liquids. Desalination. 2002;**149**:267-268. DOI: 10.1016/S0011-9164(02)00781-6

[35] Gubicza L, Nemestóthy N, Fráter T, Bélafi-Bakó K. Enzymatic esterification in ionic liquids integrated with pervaporation for water removal. Green Chemistry. 2003;**5**:236-239. DOI: 10.1039/B211342M

[36] Bélafi-Bakó K. Effect of operation conditions on the efficiency of membrane bioreactors. Desalination. 2006;**200**:511-513. DOI: 10.1016/j.desal.2006.03.416

[37] Búcsú D, Bélafi-Bakó K, Pientka Z. Two-stage membrane separation process for biohydrogen recovery. Environment Protection Engineering. 2005;**31**:33-37

[38] Búcsú D, Pientka Z, Kovács S, Bélafi-Bakó K. Biohydrogen recovery and purification by gas separation method. Desalination. 2006;**200**: 227-229. DOI: 10.1016/j.desal.2006.03.305

[39] Bélafi-Bakó K, Búcsú D, Pientka Z, Bálint B, Herbel Z, Kovács KL, et al. Integration of biohydrogen fermentation and gas separation processes to recover and enrich hydrogen. International Journal of Hydrogen Energy. 2006;**31**:1490-1495. DOI: 10.1016/j.ijhydene.2006.06.022

[40] Búcsú D, Nemestóthy N, Pientka Z, Bélafi-Bakó K. Modelling of biohydrogen production and recovery by membrane gas separation. Desalination. 2009;**240**:306-310. DOI: 10.1016/j.desal.2007.10.094

[41] Bélafi-Bakó K, Bakonyi P, Nemestóthy N, Pientka Z. Biohydrogen production in integrated system. Desalination and Water Treatment. 2010;**14**:116-118. DOI: 10.5004/dwt.2010.1017

[42] Bakonyi P, Nemestóthy N, Ramirez J, Ruiz-Filippi G, Bélafi-Bakó K. Feasibility study of gas separation membranes for biohydrogen separation. Hungarian Journal of Industrial Chemistry. 2011;**39**:331-333

[43] Bakonyi P, Nemestóthy N, Ramirez R, Ruiz-Filippi G, Bélafi-Bakó K. E. coli (XL1-BLUE) for continuous fermentation of bioH2 and its separation by polyimide membrane. International Journal of Hydrogen Energy. 2012;**37**:5623-5630. DOI: 10.1016/j.ijhydene.2012.01.009

[44] Bakonyi P, Nemestóthy N, Bélafi-Bakó K. Biohydrogen purification by membranes: An overview on the operational conditions affecting the performance of non-porous, polymeric and ionic liquid based gas separation membranes. International Journal of Hydrogen Energy. 2013;**38**:9673-9687. DOI: 10.1016/j.ijhydene.2013.05.158

[45] Bakonyi P, Nemestóthy N, Simon V, Bélafi-Bakó K. Fermentative hydrogen production in anaerobic membrane bioreactors: A review. Bioresource Technology. 2014;**156**:357-363. DOI: 10.1016/j.biortech.2014.01.079

[46] Bakonyi P, Nemestóthy N, Lankó J, Rivera R, Buitrón G, Bélafi-Bakó K. Simultaneous biohydrogen

production and purification in a double-membrane bioreactor system. International Journal of Hydrogen Energy. 2015;**40**:1690-1697. DOI: 10.1016/j.ijhydene.2014.12.002

[47] Rivera I, Buitrón G, Bakonyi P, Nemestóthy N, Bélafi-Bakó K. Hydrogen production in a microbial electrolysis cell fed with a dark fermentation effluent. Journal of Applied Electrochemistry. 2015;**45**:1223-1229. DOI: 10.1007/s10800-015-0864-6

[48] Kumar G, Bakonyi P, Kobayashi T, Xu KQ, Sivagurunathan P, Kim SH, et al. Enhancement of biofuel production via microbial augmentation: The case of dark fermentative hydrogen. Renewable and Sustainable Energy Reviews. 2016;**57**:879-891. DOI: 10.1016/j.rser.2015.12.107

[49] Rózsenberszki T, Koók L, Bakonyi P, Nemestóthy N, Logroño W, Pérez M, et al. Municipal waste liquor treatment via bioelectrochemical and fermentation (H2 + CH4) processes: Assessment of various technological sequences. Chemosphere. 2017;**171**:692-701. DOI: 10.1016/j.chemosphere.2016.12.114

[50] Bakonyi P, Kumar G, Koók L, Tóth G, Rózsenberszki T, Bélafi-Bakó K, et al. Microbial electrohydrogenesis linked to dark fermentation as integrated application for enhanced biohydrogen production: A review on process characteristics, experiences and lessons. Bioresource Technology. 2018;**251**:381-389. DOI: 10.1016/j.biortech.2017.12.064

[51] Bélafi-Bakó K, Nemestóthy N, Gubicza L. Study on application of membrane techniques in bioconversion of fumaric acid to L-malic acid. Desalination. 2004;**162**:301-306. DOI: 10.1016/S0011-9164(04)00063-3

Chapter 9

Microbial Bioremediation and Different Bioreactors Designs Applied

Memory Tekere

Abstract

Microbial remediation of pollutants involves the use of microorganisms to degrade pollutants either completely to water and carbon dioxide (for organic pollutants) or into less toxic forms. In the case of nonbiodegradable inorganic compounds, bioremediation takes the form of bioaccumulation or conversion of one toxic species to a less toxic form for example Cr(VI) is converted to less toxic (III). Bioremediation is considered an environmentally friendly way for pollution clean-up. Microbial clean up can be applied *in situ* (in place of contamination) or *ex situ* (off the site of contamination). *In situ* remediation in the natural environment is deemed slow and often times difficult to control and optimize the different parameters affecting the bioremediation. To this end, use of engineered bioreactors is preferred. Engineered bioreactors providing for optimum conditions for microbial growth and biodegradation have been developed for use in bioremediation processes to achieve the different desired remediation goals. Bioreactors in use range in mode of operation from batch, continuous, and fed batch bioreactors and are designed to optimize microbial processes in relationship to contaminated media and nature of pollutant. Designed bioreactors for bioremediation range from packed, stirred tanks, airlift, slurry phase, and partitioning phase reactors amongst others.

Keywords: bioremediation, bioreactors, pollution, microorganisms, degradation

1. Introduction

Bioremediation is a natural process that relies on microorganisms and plants and/or their derivatives (enzymes or spent biomass) to degrade or alter environmental contaminants as these organisms carry out their normal life functions [1, 2]. Bioremediation is considered an economical, versatile, efficient and eco-friendly way of dealing with environmental pollutants as compared to the physico-chemical methods [1–3]. The use of well-designed microbial bioreactors is acknowledged as an efficient way to ensure that microbial growth and processes occur in a controlled environment that provides the necessary optimum conditions [3–5]. This chapter focuses on microbial remediation in bioreactors so phytoremediation as facilitated by plants is not discussed. Several studies describe microbial remediation in designed bioreactors ranging from batch, continuous, and fed-batch operated mode which can be in different designs such as suspended carrier, slurry and fixed bed, membrane and fluidized bed reactors [4–8].

The use of microbial bioreactors in remediation is very attractive in that the bioreactors offer the advantages of providing a controlled environment where it is possible to control critical process parameters to optimize the microbial bioremediation process. Another advantage is that there is flexibility in design of the bioreactor (size and configuration) to suit application or intended purpose of the reactor [6–9]. However, bioremediation in bioreactors if operated *ex situ*, requires relocation of pollutant, a process which can involve excavation for soils and sediments, transportation and possible containment or controlled handling of the contaminated media thus making the process expensive [4–6, 8, 9]. There is a potential for exposing other environments to the contamination. Also some pretreatment of contaminated media, e.g., drying and crushing, maybe required thus adding on to the process cost [8, 9].

2. Microbial bioremediation

As defined, microbial bioremediation makes use of microorganisms and/or their derivatives (enzymes or spent biomass) to clean-up environmental contaminants [7, 9, 10]. With microorganisms, it is important to note that microorganisms are everywhere and as such pollutants in the different environmental compartments always come into contact with microorganisms [1, 2]. Microbes break down/transform pollutants via their inherent metabolic processes with or without slight pathway modifications to allow the pollutant to be channeled into the normal microbial metabolic pathway for degradation/and biotransformation. Applied bioremediation methods therefore focus on tapping the naturally occurring microbial catabolic capabilities to degrade, transform or accumulate most of the synthetic compounds such as hydrocarbons (e.g., oil), polychlorinated biphenyls (PCBs), polyaromatic hydrocarbons (PAHs), radionuclides and metals [4, 6–8]. The natural existence of a large diversity of microbial species expands the variety of chemical pollutants that are degraded or detoxified.

The advantages of microbial bioremediation are that it has public acceptance, as it is a natural process [8]. It is a low cost technology in most cases when compared to other clean-up methods for hazardous waste [2]. It can be done *in situ* and *ex situ*, instead of contaminants being transferred from one form to another or one medium to another, complete destruction of target organic pollutants is possible [8]. Notable disadvantages are that bioremediation takes relatively long to achieve treatment goals, may not be effective on all contaminants, some products of biodegradation maybe more toxic or persistent than the parent compound, specificity of the biological processes with respect to microbial populations, pollutant and environmental limitations is also a drawback and that specialized expertise are required in designing and implementing.

Bioremediation using microbial bioreactors finds application in soil, air and water environments including:

- Waste water and industrial effluent treatment

Microorganisms are the primary agents of any biological wastewater treatment. Microorganisms are already present in waste water systems and feed on complex substances in the wastewater converting them to simpler substances thus assisting in achieving the treatment. Trickling filters, membrane bioreactors, slurry phase reactors and upflow anaerobic sludge blanket bioreactors (UASB) are some of the reactors that are used in waste water and industrial effluent treatment.

- Soil and land treatment

Contaminants successfully treated include diesel fuel, fuel oils, oily sludge, wood-preserving wastes (PCP, PAHs, and creosote), coke wastes, and certain pesticides [6, 8, 9]. Soil bioremediation has proven most successful in treating petroleum hydrocarbons and other less volatile, biodegradable contaminants. Slurry phase, stirred tanks, biofilters, partitioning phase and packed microbial reactors find application in contaminated soil remediation.

- Control of air pollution

Microorganisms are used in the bioremediation of organic and inorganic air pollutants in spent gases before release or escape into the atmosphere [5, 9]. Microorganisms oxidize pollutants such as H_2S, SO_2, VOCs, and reduce pollutants such as NOx to nitrate and this assist to prevent likely environmental, health hazards and nuisances [5]. Bioscrubbers and biofilters are some of the bioreactor types often used in control of air pollution.

- Solid waste management

Microorganisms are chiefly responsible for the biodegradation of organic wastes in nature and they drive the decomposition processes that occur in landfills and composts. Anaerobic digesters are often applied mostly in the biotreatment solid waste.

2.1 Factors affecting microbial bioreactor performance

A number of issues are at play in all bioremediation technologies including when bioreactors are used. These are those that concern the contaminant, microbial community and the design, optimization and monitoring of the process [6, 8, 9]. The microbial science of bioremediation is therefore approached from many scientific frontiers: abiotic interactions (solubility, transport, sorption and photolysis), biotic interactions (taxonomic diversity, physiological, genetic and ecological interactions). In the design and operation of bioreactors in remediation, many of these factors have to be optimized and controlled for best reactor performance [5, 10–12].

Variables that affect the operation and efficiency of a microbial bioreactor relate to biotic and abiotic factors that affect microbial growth and those factors that relate to the reactor design and configuration. Factors that affect microbial growth and activities in bioreactors include; environmental factors (temperature, pH, moisture), pollutant mix, pollutant concentration, macronutrient [5, 10–12]. Factors on reactor design include; size, configuration and mode of operation.

- Environmental related factors

Environmental conditions (temperature, pH, oxygen availability/electron, and salinity) affect growth; the metabolic activities of microorganisms and to some extent the behavior of the pollutant such as solubility and volatility [11]. In any process optimization for biodegradation, it is always necessary to investigate the effects of the environmental conditions and optimize the process in relationship to all the relevant environmental conditions. Tekere et al. [13], established the optimum growth conditions with respect to pH, aeration and nutrients in the growth and degradation of pollutants by white rot fungi and found that optimized conditions result in high enzyme and degradation activities.

- Temperature

There is always a temperature range at which microorganisms grow and survive (i.e., minimum, optimum and maximum survival temperature). In addition, there is always a temperature optimum at which biochemical processes take place to achieve required bio treatment by each microorganism [13]. Extremes of temperature (too low or too high) affect both microbial growth and microbial enzyme catalyzed reactions [2]. With an increase in temperature within appropriate range, microbial metabolism increases and thus the rate of the bioremediation processes.

Increased temperatures lead to higher solubility of many chemicals, and increased fluidity and diffusion rates. For example with pollutants, such as PAHs and heavy metals, their solubility and in turn bioavailability increases with temperature [2, 7]. Temperature is thus a critical factor in the optimum operating efficiency of bioreactors to achieve best biotreatment results. Often specialized bioreactors are designed with provision for temperature control.

- pH

Similar to temperature, pH affects microbial growth and metabolic processes. pH influences microbial cell ionic properties thus microbial growth. Microorganisms have minimum, optimum and maximum pH of growth with most bacteria for example growing optimally at pH 6–7.5, though there are some which thrive best at acidic pHs (acidophiles) or at alkaline pH (alkaliphiles). Fungi generally grow at pHs lower than that of bacteria. Reactor operating pH has to be set to provide the best pH conditions for growth and enzyme activities. Behavior of pollutants is also influenced by pH thus affecting their bioremediation. For example with metals, pH affects the redox and solubility of metals, different forms and valence have different effects on microorganisms [14]. Metal solubility increases with a decrease in medium pH and alkaline pH favor metal ion precipitation. Often lower pH values are required for metal attachment to the microbial cell surface [7, 14]. Microorganisms that produce acids result in increased solubility of the metal ions [10]. To provide for best pH conditions, buffers are used in media formulations, acids and bases can be added during the bioreactor process [13].

- Nutrients

Nutrients are required for growth and metabolism of the microorganisms. Several elements are required in biosynthesis and energy production. Carbon is the most basic element of living forms and is needed in greater quantities than other elements. Other elements that are important in ensuring a balanced nutritional bioreactor environment depending on the type of microorganism include hydrogen, oxygen, nitrogen, sulfur, phosphorus, iron, calcium and magnesium [10, 11]. All necessary macro- and micro- nutrients requirements are provided in reactor media. Microorganisms can use the pollutants they are degrading as primary energy sources or a primary source of energy is provided to the microorganism in the case of co metabolism of the pollutants.

- Moisture

Water is required to support microbial growth and catalysis. Cellular chemical reactions occur in aqueous conditions and water is required to ensure the correct osmotic pressure is maintained for microbial growth. The amount of water available for microbial growth is called (aW). Most microorganisms grow at water activities of 0.98 or higher, osmotolerant species can however grow at a range of low aW [11].

- Electron acceptors

The presence of electron acceptors, e.g., oxygen in aerobic microbes and NO_3^{1-}, SO_4^{2-} and Fe (III) oxides in case of anaerobic microbes, also affects the biodegradation processes.

- Reactor design related factors

Bioreactors have to provide for the best conditions for microbial growth and biochemical process to occur. The reactor size, configuration and mode of operation are key reactor design factors. The reactor should provide favorable physical, biological and the combined physical-chemical conditions for the best biological remediation processes to be achieved. In designing the bioreactor, favorable physical conditions for transport of gases and liquids and solids over time that ensure that the physical entity of the bioreactor is favorably adapted to the biological system that performs the bioreactions are required [12, 15]. On the other hand there is need to ensure that the biophysical and biochemical events taking operate at optimum levels under real situation application.

Polluted samples for remediation can be fed into the reactor either as dry or slurry matter [9]. Pollutants with hydrophobic properties are often unavailable for microbial degradation, particularly if they are bound to soil matrix [7]. Their degradation is therefore limited by their transfer to liquid [4]. Minimizing mass transfer resistance was found to be a key factor in the degradation of hexachlorocyclohexane (HCH) in slurry batch bioreactors [4].

Despite the rapid development of bioreactors due to their widespread use in biotechnology, the aspects of maintaining stability and rates of bioprocesses are still areas to be addressed. Poor bioreactor construction and design, leading to inadequate mixing, may jeopardize the stability and performance of the process [15]. Mixing prevents thermal stratification, help maintain uniform conditions in the reactor, ensure good contact between microbial culture and media reactants. The importance of mixing in bioreactor cannot be over emphasized, poor mixing affect microbial process efficiency.

Hydraulic retention times (HRT) required to achieve the necessary remediation goals in the bioreactor have to be determined and optimized. Longer HRTs result in poor substrate loading which diminishes the microbial population, whereas shorter ones do not allow microorganisms to effectively degrade the pollutant and can result in microbial wash out from the system [16].

- Organism related factors

Organism related factors include population density, composition, inter and intraspecific interaction. Microbes are the most diverse forms of life and have developed a wide range of metabolic pathways that enable them to cope under the varying ecological conditions including exposure to xenobiotics. A whole range of environments ranging from aerobic, anaerobic, acidic, alkaline, and low to high temperature have been utilized as sources of microorganisms for bioremediation [13]. Only certain species of bacteria and fungi have proven their ability as potent pollutant degraders [13]. In the natural environment degradation of pollutants is often achieved through complex microbial population interactions. Single or mixed microbial cultures are used for pollutant remediation in bioreactors. In the event where bioagumentation is applied the introduced organisms need to be able to co-exist with indigenous residents.

Different microorganisms often have different metabolic capabilities, to this extend the evaluation of several strains of different microbial players have to be investigated in order to come up with the best degraders [13]. In screening and comparison of the bio-degradation of PAHs by white rot fungi [17], found out that newly screened white rot fungi strains had higher or comparable degradation capacity to the model well applauded *P. chrysosporium*, and these strains did not accumulate the metabolite quinone which accumulates as a dead end metabolite in *P. chrysosporium*.

Polluted environments provide sources of microorganisms resistant or acclimatized to the pollutant [18]. However microorganisms that are known to have certain inherent physiological characteristic, e.g., metabolism of known substrate with structural similarity to xenobiotics of interest and/or adaptation to certain environmental conditions can be selected. This is the case in several studies that used microorganisms for pollutant degradation [11, 17–19].

- Pollutant related factors

Factors that affect bioremediation in bioreactors that are related to the pollutant include: nature of pollutant, i.e., the physical and chemical properties including solubility, volatility, molecular complexity, concentration and toxicity. Investigations for most pollutant biodegradation have centered on how different concentrations, mixed pollutants, solubility and molecular structure can affect microbial bioremediation [17, 20]. In the case of PAHs, degradation decreases in the order alkane> branched chain alkanes>low molecular weight aromatics> cycloalkanes [17]. It should be noted however that some pollutants are resistant to biodegradation (recalcitrant, i.e., resistant to degradation) they are degraded at very low pace even if the right microbial population and conditions are present.

3. Microbial bioreactors in bioremediation

Several laboratory, and pilot bioremediation studies have been done using microbial (fungi and bacteria) bioreactors [6, 8, 17, 18, 20]. Bioreactor technologies may offer effective means for treatment of many contaminants in groundwater, soil and air [4, 5, 7, 12]. The bioreactor type of choice for any application should be easy to operate and maintain for the selected purpose and application. **Table 1** presents some of the studies that involved the use of bioreactors in bioremediation. Flexibility to design bioreactor tailor made for different processes and remediation applications makes the use of bioreactors in bioremediation attractive [9]. The design should accommodate high biomass from cell growth, supply of necessary nutrients and also removal of waste components from the system. A description of some bioreactor types and their application is given in Sections 3.1–3.7.

3.1 Slurry phase bioreactors

Slurry phase bioreactors, as the name implies treats polluted media that is within a slurry phase. Alternative names are bio-slurry reactors and slurry phase biological treatment. Slurry bioreactors offer an *ex situ* environmentally friendly way for remediating mostly soils and sediments from petrochemical hydrocarbons, tars, creosotes, chlorinated solvents, herbicides, pesticides and explosives or when a solid substrate that is formulated into a slurry is used [4, 6, 25, 26]. Hydrophobic nature

Bioreactor type	Application details	Reference(s)
Packed bed	Different fungi and bacteria used for remediation of organochlorine pesticides, PAHs, pharmaceuticals, amines, and textile dyes. Packing material varied from organic material (sawdust, wood chips) to inert solid materials (polyurethane foam, poraver stones); chlorinated aliphatic compounds	[14, 17, 21–23]
Fluidised bed	Treatment of pharmaceuticals using fungi	[20, 22]
Two-phase partitioning	Benzene biodegradation by cow dung microflora	[24]
Slurry phase	Bacterial and fungal remediation of soil from VOC, organochlorines, PAHs, 2,4-dichlorophenoxyacetic acid	[4, 6, 25, 26]
Suspended carrier	Fungi used for remediation of organochlorine pesticides, PAHs, textile dyes.	[21]
Up-flow anaerobic stage reactor (UASR)	Bacterial degradation of tylosin	[16]
Membrane bioreactor	Textile dye in waste water; pharmaceuticals, 1,2-dichloroethane, 1,2-dichlorobenzene and 2-chlorophenol groundwater; metal recovery	[27–31]
Air lift	Textile dye effluent decolorization by fungi, olive mill effluent, cellulose industry bleaching effluent	[15, 32, 33]
Biotrickling filter	Municipal waste water, brewery waste, olive oil mill waste water, VOC contaminated air	[32, 34]
Upflow Anaerobic Sludge Blanket	Potato waste water, BTEX	[9, 35]
Sequence batch reactor	Nanosilver, Nanofullerenes	[36]
Continuous flow Bioreactor	PCP and creosote by some Pseudomonas species	[37]
Nonisothermal bioreactors	Degradation of phenol by fungal laccase	[38]
Continuously stirred tank bioreactor (CSTR)	For hydrocarbon-rich industrial wastewater effluents by mixed microbial cultures, petroleum hydrocarbon	[7, 8]

Table 1.
Studies that involved the use of bioreactors in bioremediation.

of most persistent chemicals makes them sorb to soil or sediments and not easily accessible for biodegradation.

Operation of the slurry reactor can be in batch, semi-continuous and continuous mode, with the batch process being the most common one [6, 26]. **Figure 1** shows an illustration of a simplified slurry reactor. Water is mixed with the contaminated solid matrix in suitable ratios and this enhances contact between microorganisms, pollutant, media and oxygen. Pollutants that are solubilized become more bioavailable. **Table 2** shows some of the studies that have involved the use of slurry phase bioreactors in bioremediation.

3.2 Partitioning bioreactors

Partitioning bioreactors are used in bioremediation when two phases need to be achieved, e.g., such as for organic solvents or water immiscible compounds in

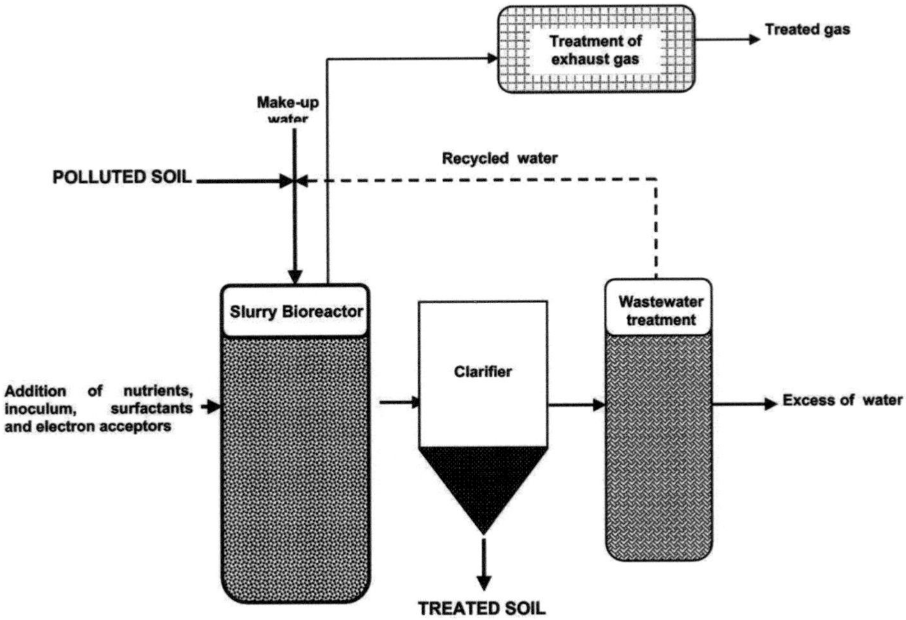

Figure 1.
Simplified slurry reactor [26].

aqueous solutions. Reactors are designed with the aqueous and organic phase, and can be single or multiphased [24]. With toxic hazardous waste, toxicity to degrading microorganisms is a problem. In partitioning bioreactors, there is a two-phase system where a water immiscible and biocompatible organic solvent is allowed to float on the surface of a cell containing aqueous phase [45]. This means that high amounts of hazardous waste dissolved in a solvent can be added to the reactor without the microorganism experiencing inhibitory concentrations of the pollutant [24, 45, 46]. A rigorous process involving selection of the solvent, taking into consideration the biological, physical, operational, environmental and economic factors is necessary in developing an efficient partitioning biotreatment system. Partitioning reactors find application in the remediation of toxic compounds from petrochemical industry such as benzene as well as VOC in waste gases of many industrial processes [45, 47, 48]. Angelucci et al. [49], successfully tested a continuous two-phase-partitioning reactor in the treatment of tannery wastewater. Several other studies involving phase partitioning bioreactors are described [24, 45–50].

3.3 Stirred tank bioreactors

A continuous stirred tank bioreactor consists of a cylindrical vessel with motor driven central shaft that supports one or more agitators (impellers). Stirred tank bioreactors are the predominantly used design for submerged cultures. Stirred tank bioreactors are mechanically agitated where the stirrers are the main gas-dispersing tools and provide high values of mass transfer rates coupled with excellent mixing. Advantages of the STR include the efficient gas transfer to growing cells, good mixing of the contents and flexible operating conditions, besides the commercial availability of the bioreactors. The main shortcoming of the stirred tank bioreactor is its mechanical agitation which requires energy and stirring can cause shear strain on microbial cells.

Gargouri et al. [7] evaluated the use of a continuously stirred tank bioreactor (CSTR) in the treatment of hydrocarbon-rich industrial wastewaters and achieved

Pollutant	Microorganism(s)	Bioremediation details	Reference(s)
Petroleum hydrocarbons in oil sludge	Indigenous microbial consortium	24% biodegradation of Total Petroleum Hydrocarbon in oily waste	[39]
2,4,6-trinitrotoluene (TNT)	Mixed soil bacteria under anoxic/microaerophilic conditions	99% of 10,000 mg kg^{-1} was degraded in 82 days under co-metabolism with molasses	[40]
PAHs in creosote	Degradation by *Pseudomonas fluorescens, Pseudomonas stutzeri,* and an *Alcaligenes* species	93.4% of creosote degraded in 12 weeks	[41]
Explosives 2,4,6-trinitrotoluene (TNT) and 2,4,6-trinitrobenzene (TNB)	Selected Gram positive bacterial isolates	Complete removal of the explosive after 80 days	[42]
Hexachlorocyclohexane (HCH)	White rot fungi *Bjerkandera adusta*	Maximal degradations of 94.5, 78.5 and 66.1% were attained after 30 days for the-HCH isomers, respectively	[4]
High molecular weight PAH in soil	PAH-degrading consortium	Pyrene degraded at 19 mg L^{-1} day^{-1}, chrysene and benzo[*a*]pyrene respectively at 3.5 and 0.94 mg L^{-1} day^{-1}.	[43]
Chlorpyrifos	Enriched indigenous soil microorganism	Degradation of 48% in aerobic and 31% in anaerobic soil slurries	[44]

Table 2.
Some examples of remediation studies in slurry phase bioreactor.

successful bioremediation using an acclimatized microbial consortium. The residual total petroleum hydrocarbon (TPH) decreased from 320 –8 mg TPH l^{-1}. The reactor used is shown in **Figure 2**. Bi [51], applied a continuously stirred tank reactor for bioremediation of ethanol, toluene and benzyl alcohol by *P. putida*.

3.4 Biofilters

A basic biofilter bioreactor consist of a large media bed where pollutants are passed through and get degraded by the microorganisms. Biofilters are amongst the oldest environmental bioremediation techniques. Biofilters are used mostly in waste water treatment as well as in the control of air pollution [34, 52, 53]. A number of materials are used for bed media such as peat, composted yard waste, bark, coarse soil, gravel or plastic shapes. A typical example of a biofilter is the trickling filter which finds extensive application in the treatment of different liquid effluents or waste waters or waste that is constituted into liquid. A trickling filter is usually a round, vertical tank that contains a support rack and is filled with aggregate, ceramic or plastic media and in the middle of the tank is a vertical pipe that has a rotary connection with spray nozzles on the top end [34]. A spray arm is attached to the rotary connection and has spray nozzles installed along its length for

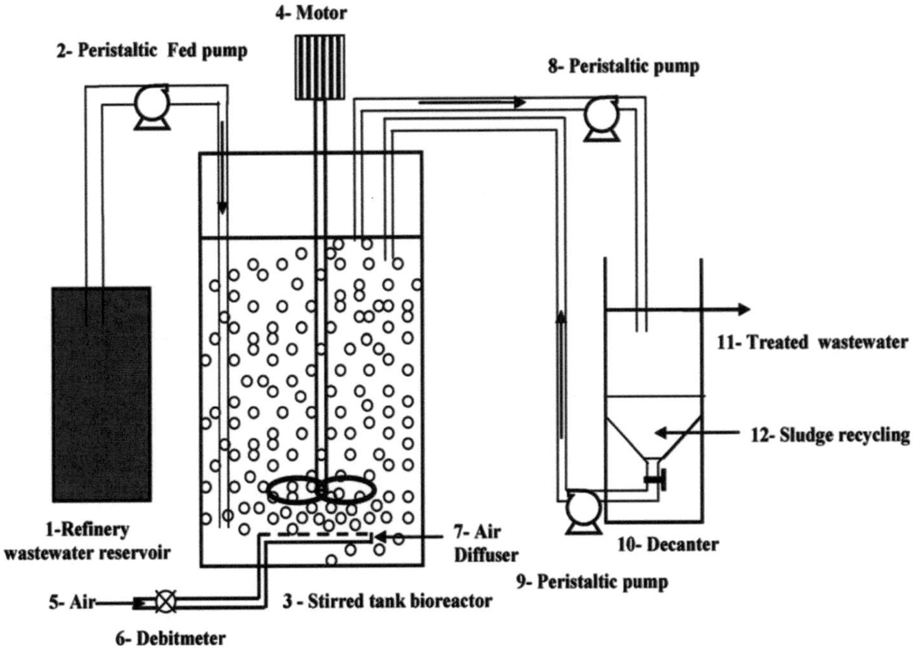

Figure 2.
Schematic diagram of the aerobic continuously stirred tank bioreactor (CSTR) used for continuous experiments [7].

distribution of the waste water. Microorganisms grow in biofilm forms on the packing material surface and are responsible for the degradation of the pollutants from the effluent. Schmidt and Anderson [34] described the use of a trickling biofilter in the removal of high concentrations of 1-butanol from contaminated air. The potential application of the biotrickling filter in industrial off gas treatment was evaluated in the removal of high concentrations of 1-butanol from contaminated air with efficiency exceeding 80% for butanol concentrations of 0.4–1.2 g m^{-3} [34]. The laboratory-scale perlite-packed biotrickling filter was operated for 60 days and demonstrated effective and efficient removal of butanol concentrations up to 4.65 g m^{-3} with a maximum elimination capacity of 100 g m^{-3} h^{-1} [34].

3.5 Packed bed bioreactors

Packed bed bioreactor systems provide for microbial growth on fixed film substrata. In order to obtain compact reactors and ensure greater treatment reliability, fixed film reactors are used. They offer the advantage that dilute aqueous solutions can be remediated at high biomass without the need to separate biomass and the treated effluent [13, 54]. In packed bed biofilm biotreatment processes, unlike suspension cultures there is no need to incorporate special measures such as centrifugation and membrane filters to retain the biomass. This feature makes the use of packed bed reactors particularly appropriate in bioreactors systems where large substrate—flow through is required. The concentration of cells in a given volume may be increased, a factor that leads to enhanced efficiency/productivity of the bioreactor and decreased volume of bioreactors [55]. While high biomass concentrations can be easily maintained, the medium to biofilm mass transfer of substrate is the rate limiting process in packed bed bioreactors [54, 56]. Within the biofilm there are considerable differences in the microorganisms' microenvironment, depending on the distance from the surface of the biofilm [54]. Substrates such as oxygen, carbon and nitrogen sources have to cross the biofilm—liquid interface by diffusion, thus a diffusion gradient occurs. To calculate the kinetics of conversion in the biofilm processes, two important processes that occur in the system are considered and these are (i) transport of solutes over the biofilm and (ii) combined reactions and diffusion inside the biofilm [54]. In the packed bed reactors, development of excess microbial biomass also occurs leading to hydraulic channeling or loss of interstitial fluid volume. To overcome the severe constraints of hydraulic hold up within the interior of the reactor extra-capillary space transverse flow bioreactors were developed [57].

Selection of suitable substances as packing materials is an important consideration. Materials that have been used include nylon web, polyurethane foam, silicone tubing, sintered glass, porous ceramics, propylene, stainless steel, agarose and agar gel beads [58–67]. The ideal support should be chemically inert in physiological growth medium, rigid and porous to facilitate mycelial attachment and re-usable after removal of the fungus. **Figure 4** shows a Simplified diagram of a laboratory based packed bed bioreactor. Examples of remediation studies in packed bed reactors are given in **Table 3**.

3.6 Airlift bioreactors

Airlift bioreactors can provide an attractive treatment alternative for treatment of gaseous or volatile air pollutants. Frequently, the most limiting factor in the performance of these reactors is that they are susceptible to being limited by gas-liquid mass transfer and by poor mixing of the liquid phase, particularly when they are operating at high cell densities [68, 69]. The bioreactor performance is dependent on the pumping (injection) of air and the liquid circulation. The airlift bioreactor can have a forced

Support	Experimental study details	References
Polyurethane foam	Anaerobic fixed film horizontal flow bench scale reactor. Benzene, toluene, ethylbenzene, and xylene, BTEX removal with efficiency of 75–99% in 11.4 hrs	[58]
Laterite stones	Microbial consortium anaerobic degradation of textile azo dyes, 61.7% degradation of 55 $\mu g\ mL^{-1}$ of simulated effluent dye.	[59]
Coconut shell bio-char	Congo red dye degradation in batch and continuous packed bed bioreactors by *Brevibacillus parabrevis.* A 95.71% removal of in 6 days of 150 ppm dye.	[60]
Polyurethane foam	Bacterial degradation of malathion in batch and continuous packed bed bioreactors, removal at 89% for up to 145.4 $mg\ L^{-1}\ day^{-1}$	[61]
Wire Mesh	Fungal degradation of textile effluent	[62]
Wood chips	Chlorophenol degradation by *Phanerochaete chrysosporium*	[63]
Sugarcane bagasse	Degradation of dyes and industrial effluents by *Garnoderma weberianum* B-18 immobilized in a lab-scale packed-bed bioreactor. 55–98% for different dyes tested	[64]
Celite	Perchlorate-Contaminated groundwater 800 $\mu g\ L^{-1}$ reduced to less than 4 μ^{-1} at 0.3 h retention time	[66]
Polyurethane foam	Biodegradation of an actual petroleum wastewater by an immobilized biomass of *Bacillus cereus*	[66]
Polyurethane foam and alginate beads	Benzene biodegradation Bacillus sp. M3 at 84 in alginate beads and 90% on polyurethane foam within 9 days	[67]

Table 3.
Some examples of remediation studies in packed bed reactors.

flow in an internal or external loop as shown in **Figure 5**. Specific volatile organic chemicals may be completely degraded by a microorganism at normal temperature and pressure without producing a second polluted byproduct [70]. Nikakhtari and Hill [68], applied and External Loop Airlift Bioreactor with a small amount (99% porosity) of a stainless steel mesh packing inserted in the riser section for bioremediation of a phenol polluted air stream. Phenol removal of 100% was achieved using the bacterium *Pseudomonas putida*, and at a phenol loading rate of 22,160 mg h^{-1} m^{-3}, thus demonstrating the novelty and potential VOCs bioremediation application of the reactor at high loading rates. **Figure 5** presents a schematic diagram of airlift bioreactor. Several other studies involving the use of airlift bioreactors [19, 69–71].

3.7 Membrane bioreactor

Membrane bioreactors (MBR) combine the use of a membrane that forms a filtration system and the biological process. The membrane provides a physical barrier that separates the liquid from the solid and ensures retention of the solids and good quality effluent. The quality of the treated effluent from the membrane bioreactor is of high quality than that achieved by employing other techniques, enabling optimal functioning of the secondary treatment system [72, 73]. MBR offer the advantages that often smaller tank size is used and filtration function of the membrane ensures that solids are separated from treated effluent. Membrane fouling has been recognized however as a major drawback in the application of membrane bioreactors in bioremediation. Also membranes are often expensive thus making the process costly. Development

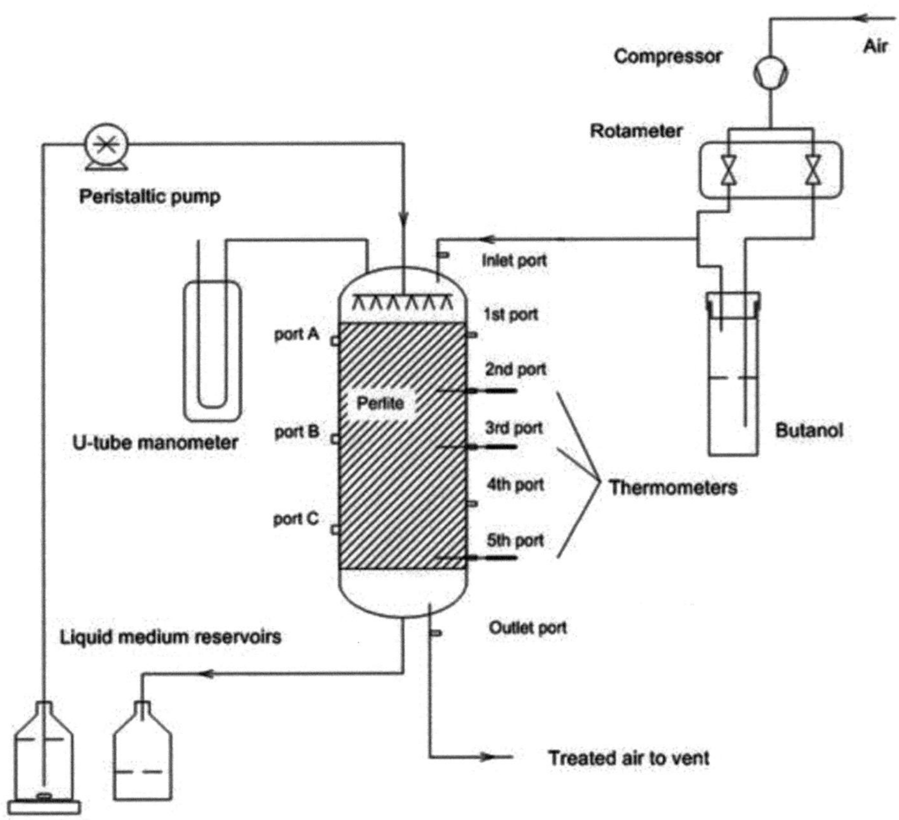

Figure 3.
Schematic diagram of biotrickling filter [34].

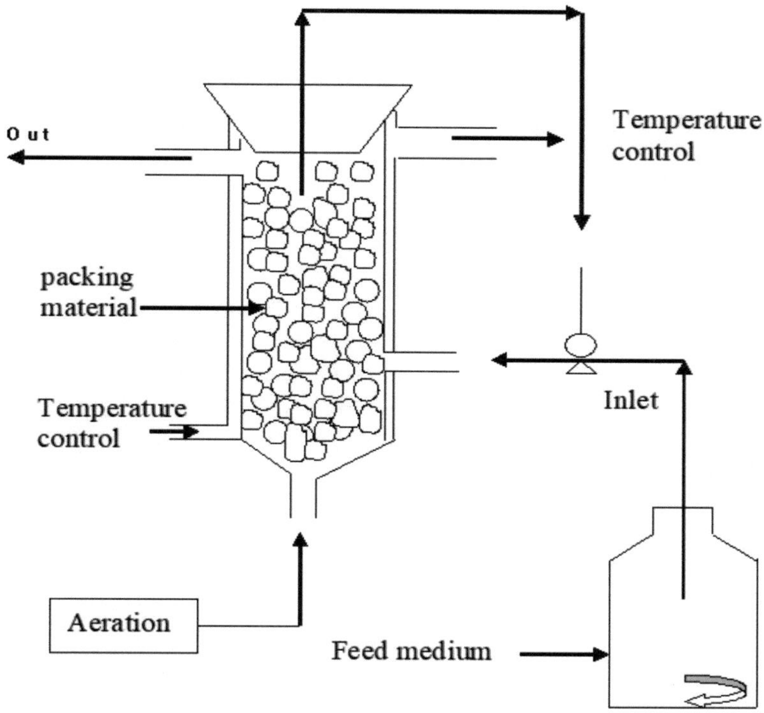

Figure 4.
Simplified diagram of a laboratory based packed bed bioreactor [21].

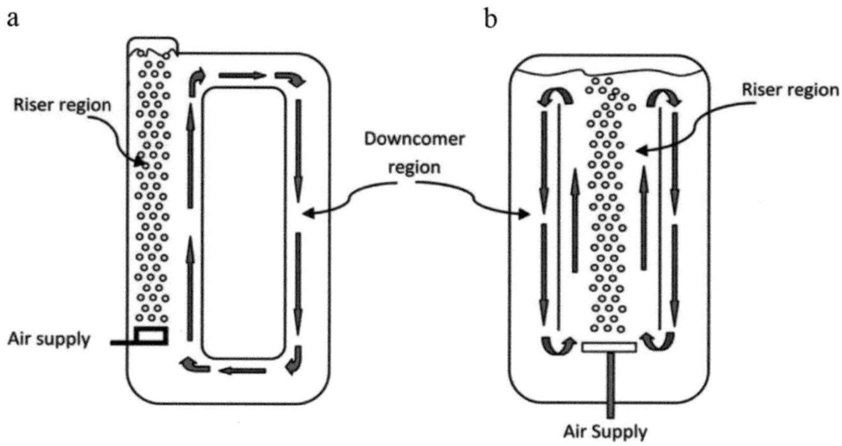

Figure 5.
Schematic diagram of airlift bioreactor with (a) external recirculation and (b) internal recirculation [15].

of low cost membrane filters is an ongoing feature in the science of MBR [72]. MBR reactors have been used in the biological treatment of domestic and industrial waste water. MBR have been evaluated in the remediation of pentachlorophenol in concentration ranges that occur in waste water [73], textile waste water [27], 1,2-dichloroethane, 1,2-dichlorobenzene and 2-chlorophenol [30].

3.8 Other bioreactors in bioremediation

Due to flexibility in bioreactor designs, the configuration of reactors is numerous. While an effort has been made here to describe some of the common bioreactors used for different bioremediation applications, several other bioreactor types have not been discussed. These include the UASB which find major application in anaerobic digestion of waste waters as well as solid wastes, bio-scrubbers which are applied in off gas air pollution control, continuous stirred tank reactors as well as rotating contactor reactors.

4. Conclusions

It is evident that a wide range of microbial bioreactors have been developed and evaluated in the bioremediation of a wide range of pollutants in water, air and soil. Also a wide range of pollutants in physical and chemical properties are amenable to microbial degradation. Very diverse microbial species have the capability of pollutant degradation naturally and the use of well-developed optimized microbial bioreactors ensure improved rates of degradation when compared to degradation that happens *in situ* in the environment under natural environmental conditions.

Conflict of interest

No conflict of interest is declared.

Author details

Memory Tekere
Environmental Science Department, University of South Africa, Johannesburg, South Africa

*Address all correspondence to: tekerm@unisa.ac.za

References

[1] Kumar A, Bisht BS, Joshi VD, Dhewa T. Review on bioremediation of polluted environment: A management tool. International Journal of Environmental Sciences. 2011;**1**(6):1079

[2] Sharma S. Bioremediation: Features, strategies and applications. Asian Journal of Pharmacy and Life Science. 2012;**2231**:4423

[3] Jeon CO, Madsen EL. In situ microbial metabolism of aromatic-hydrocarbon environmental pollutants. Current Opinion in Biotechnology. 2013;**24**(3):474-481

[4] Quintero JC, Lu-Chau TA, Moreira MT, Feijoo G, Lema JM. Bioremediation of HCH present in soil by the white-rot fungus Bjerkandera adusta in a slurry batch bioreactor. International Biodeterioration & Biodegradation. 2007;**60**(4):319-326

[5] U.S. Environmental Protection Agency. Using bioreactors to control air pollution—U.S. Environmental Protection Agency, EPA-456/R-03-003. 2003. Available from: http://www.epa.gov/ttn/catc [Accessed: Nov 23, 2018]

[6] Pino-Herrera DO, Pechaud Y, Huguenot D, Esposito G, Van Hullebusch ED, Oturan MA. Removal mechanisms in aerobic slurry bioreactors for remediation of soils and sediments polluted with hydrophobic organic compounds: An overview. Journal of Hazardous Materials. 2017;**339**:427-449

[7] Gargouri B, Karray F, Mhiri N, Aloui F, Sayadi S. Application of a continuously stirred tank bioreactor (CSTR) for bioremediation of hydrocarbon-rich industrial wastewater effluents. Journal of Hazardous Materials. 2011;**189**(1-2):427-434

[8] Chikere CB, Chikere BO, Okpokwasili GC. Bioreactor-based bioremediation of hydrocarbon-polluted Niger Delta marine sediment, Nigeria. 3 Biotech. 2012;**2**(1):53-66

[9] Azubuike CC, Chikere CB, Okpokwasili GC. Bioremediation techniques–classification based on site of application: Principles, advantages, limitations and prospects. World Journal of Microbiology and Biotechnology. 2016;**32**(11):180

[10] Srivastava J, Naraian R, Kalra SJ, Chandra H. Advances in microbial bioremediation and the factors influencing the process. International Journal of Environmental Science and Technology. 2014;**11**(6):1787-1800

[11] Naik MG, Duraphe MD. Review paper on-parameters affecting bioremediation. International Journal of Life Science and Pharma Research. 2012;**2**(3):L77-L80

[12] Mandenius CF. Bioreactors: Design, Operation and Novel Applications. Germany: John Wiley & Sons; 2016

[13] Tekere M, Mswaka AY, Zvauya R, Read JS. Growth, dye degradation and ligninolytic activity studies on Zimbabwean white rot fungi. Enzyme and Microbial Technology. 2001;**28**(4-5):420-426

[14] Liu SH, Zeng GM, Niu QY, Liu Y, Zhou L, Jiang LH, et al. Bioremediation mechanisms of combined pollution of PAHs and heavy metals by bacteria and fungi: A mini review. Bioresource Technology. 2017;**224**:25-33

[15] Mahmood KA, Wilkinson SJ, Zimmerman WB. Airlift bioreactor for biological applications with microbubble mediated transport processes. Chemical Engineering Science. 2015;**137**:243-253

[16] Chelliapan S, Wilby T, Yuzir A, Sallis PJ. Influence of organic loading on the performance and microbial community

structure of an anaerobic stage reactor treating pharmaceutical wastewater. Desalination. 2011;**271**(1-3):257-264

[17] Tekere M, Read JS, Mattiasson B. Polycyclic aromatic hydrocarbon biodegradation in extracellular fluids and static batch cultures of selected sub-tropical white rot fungi. Journal of Biotechnology. 2005;**115**(4):367-377

[18] Lladó S, Gràcia E, Solanas AM, Viñas M. Fungal and bacterial microbial community assessment during bioremediation assays in an aged creosote-polluted soil. Soil Biology and Biochemistry. 2013;**67**:114-123

[19] Ryan DR, Leukes WD, Burton SG. Fungal bioremediation of phenolic wastewaters in an airlift reactor. Biotechnology Progress. 2005;**21**(4):1068-1074

[20] Gros M, Cruz-Morato C, Marco-Urrea E, Longrée P, Singer H, Sarrà M, et al. Biodegradation of the X-ray contrast agent iopromide and the fluoroquinolone antibiotic ofloxacin by the white rot fungus Trametes versicolor in hospital wastewaters and identification of degradation products. Water Research. 2014;**60**:228-241

[21] Tekere M, Read JS, Mattiasson B. Polycyclic aromatic hydrocarbon biodegradation by a subtropical white rot fungus in packed bed and suspended carrier bioreactor systems. Environmental technology. 2007;**28**(6):683-691

[22] Svobodová K, Petráčková D, Kozická B, Halada P, Novotný Č. Mutual interactions of *Pleurotus ostreatus* with bacteria of activated sludge in solid-bed bioreactors. World Journal of Microbiology and Biotechnology. 2016;**32**(6):94

[23] Juarez-Ramirez C, Galíndez-Mayer J, Ruiz-Ordaz N, Ramos-Monroy O, Santoyo-Tepole F, Poggi-Varaldo H. Steady-state inhibition model for the biodegradation of sulfonated amines in a packed bed reactor. New Biotechnology. 2015;**32**(3):379-386

[24] Singh D, Fulekar MH. Benzene bioremediation using cow dung microflora in two phase partitioning bioreactor. Journal of Hazardous Materials. 2010;**175**(1-3):336-343

[25] Plangklang P, Reungsang A. Bioaugmentation of carbofuran by Burkholderia cepacia PCL3 in a bioslurry phase sequencing batch reactor. Process Biochemistry. 2010;**45**(2):230-238

[26] Robles-González IV, Fava F, Poggi-Varaldo HM. A review on slurry bioreactors for bioremediation of soils and sediments. Microbial Cell Factories. 2008;**7**(1):5

[27] Hossain K, Ismail N. Bioremediation and detoxification of pulp and paper mill effluent: A review. Research Journal of Environmental Toxicology. 2015;**9**(3):113

[28] Couto CF, Lange LC, Amaral MC. A critical review on membrane separation processes applied to remove pharmaceutically active compounds from water and wastewater. Journal of Water Process Engineering. 2018;**26**:156-175

[29] Jegatheesan V, Pramanik BK, Chen J, Navaratna D, Chang CY, Shu L. Treatment of textile wastewater with membrane bioreactor: A critical review. Bioresource Technology. 2016;**204**:202-212

[30] Carucci A, Manconi I, Manigas L. Use of membrane bioreactors for the bioremediation of chlorinated compounds polluted groundwater. Water Science and Technology. 2007;**55**(10):209-216

[31] Mack C, Burgess JE, Duncan JR. Membrane bioreactors for metal

recovery from wastewater: A review. Water SA. 2004;**30**(4):521-532

[32] Olivieri G, Russo ME, Giardina P, Marzocchella A, Sannia G, Salatino P. Strategies for dephenolization of raw olive mill wastewater by means of *Pleurotus ostreatus*. Journal of Industrial Microbiology & Biotechnology. 2012;**39**(5):719-729

[33] Souza ÉS, Souza JV, Silva FT, Paiva TC. Treatment of an ECF bleaching effluent with white-rot fungi in an air-lift bioreactor. Environmental Earth Sciences. 2014;**72**(4):1289-1294

[34] Schmidt T, Anderson WA. Biotrickling filtration of air contaminated with 1-butanol. Environments. 2017;**4**(3):57

[35] Manhokwe S, Parawira W, Tekere M. An evaluation of a mesophilic reactor for treating wastewater from a Zimbabwean potato-processing plant. African Journal of Environmental Science and Technology. 2009;**3**(4):091-096

[36] Yang Y, Wang Y, Hristovski K, Westerhoff P. Simultaneous removal of nanosilver and fullerene in sequencing batch reactors for biological wastewater treatment. Chemosphere. 2015;**125**:115-121

[37] Mueller JG, Lantz SE, Ross D, Colvin RJ, Middaugh DP, Pritchard PH. Strategy using bioreactors and specially selected microorganisms for bioremediation of groundwater contaminated with creosote and pentachlorophenol. Environmental Science & Technology. 1993;**27**(4):691-698

[38] Georgieva S, Godjevargova T, Portaccio M, Lepore M, Mita DG. Advantages in using non-isothermal bioreactors in bioremediation of water polluted by phenol by means of immobilized laccase from *Rhus vernicifera*. Journal of Molecular Catalysis B: Enzymatic. 2008;**55**(3-4):177-184

[39] Machin-Ramirez C, Okoh AI, Morales D, Mayolo-Deloisa K, Quintero R, Trejo-Hernández MR. Slurry-phase biodegradation of weathered oily sludge waste. Chemosphere. 2008;**70**(4):737-744

[40] Clark B, Boopathy R. Evaluation of bioremediation methods for the treatment of soil contaminated with explosives in Louisiana Army Ammunition Plant, Minden, Louisiana. Journal of Hazardous Materials. 2007;**143**(3):643-648

[41] Lewis RF. Site demonstration of slurry-phase biodegradation of PAH contaminated soil. Air & Waste. 1993;**43**(4):503-508

[42] Fuller ME, Manning JF Jr. Microbiological changes during bioremediation of explosives-contaminated soils in laboratory and pilot-scale bioslurry reactors. Bioresource Technology. 2004;**91**(2):123-133

[43] Villemur R, Deziel E, Benachenhou A, Marcoux J, Gauthier E, Lépine F, et al. Two-liquid-phase slurry bioreactors to enhance the degradation of high-molecular-weight polycyclic aromatic hydrocarbons in soil. Biotechnology Progress. 2000;**16**(6):966-972

[44] Tiwari MK, Guha S. Kinetics of biotransformation of chlorpyrifos in aqueous and soil slurry environments. Water Research. 2014;**51**:73-85

[45] Daugulis AJ. Two-phase partitioning bioreactors: A new technology platform for destroying xenobiotics. Trends in Biotechnology. 2001;**19**(11):457-462

[46] Davidson CT, Daugulis AJ. Addressing biofilter limitations: A two-phase partitioning bioreactor

process for the treatment of benzene and toluene contaminated gas streams. Biodegradation. 2003;**14**(6):415-421

[47] Dorado AD, Dumont E, Muñoz R, Quijano G. A novel mathematical approach for the understanding and optimization of two-phase partitioning bioreactors devoted to air pollution control. Chemical Engineering Journal. 2015;**263**:239-248

[48] Muñoz R, Daugulis AJ, Hernández M, Quijano G. Recent advances in two-phase partitioning bioreactors for the treatment of volatile organic compounds. Biotechnology Advances. 2012;**30**(6):1707-1720

[49] Angelucci DM, Stazi V, Daugulis AJ, Tomei MC. Treatment of synthetic tannery wastewater in a continuous two-phase partitioning bioreactor: Biodegradation of the organic fraction and chromium separation. Journal of Cleaner Production. 2017;**152**:321-329

[50] Tomei MC, Angelucci DM, Daugulis AJ. Towards a continuous two-phase partitioning bioreactor for xenobiotic removal. Journal of Hazardous Materials. 2016;**317**:403-415

[51] Bi Y. Bioremediation of volatile organic compounds in a continuous stirred tank bioreactor (doctoral dissertation). Saskatoon: University of Saskatchewan; 2005

[52] Lebrero R, Gondim AC, Pérez R, García-Encina PA, Muñoz R. Comparative assessment of a biofilter, a biotrickling filter and a hollow fiber membrane bioreactor for odor treatment in wastewater treatment plants. Water Research. 2014;**49**:339-350

[53] Rabbani KA, Charles W, Kayaalp A, Cord-Ruwisch R, Ho G. Pilot-scale biofilter for the simultaneous removal of hydrogen sulphide and ammonia at a wastewater treatment plant. Biochemical Engineering Journal. 2016;**107**:1-10

[54] Van Loosdrecht MC, Heijnen SJ. Biofilm bioreactors for waste-water treatment. Trends in Biotechnology. 1993;**11**(4):117-121

[55] Bisping B, Rehm HJ. Multistep reactions with immobilized microorganisms. Biotechnology and Applied Biochemistry. 1988;**10**(2):87-98

[56] Guieysse B, Bernhoft I, Andersson BE, Henrysson T, Olsson S, Mattiasson B. Degradation of acenaphthene, phenanthrene and pyrene in a packed-bed biofilm reactor. Applied Microbiology and Biotechnology. 2000;**54**(6):826-831

[57] Burton SG, Boshoff A, Edwards W, Jacobs EP, Lfukes WD, Rose PD, et al. Membrane Based Biotechnological System for the Treatment of Organic Pollutants. South Africa: Water Research Commission; 1998

[58] de Nardi IR, Ribeiro R, Zaiat M, Foresti E. Anaerobic packed-bed reactor for bioremediation of gasoline-contaminated aquifers. Process Biochemistry. 2005;**40**(2):587-592

[59] Senan RC, Shaffiqu TS, Roy JJ, Abraham TE. Aerobic degradation of a mixture of Azo dyes in a packed bed reactor having bacteria-coated laterite pebbles. Biotechnology Progress. 2003;**19**(2):647-651

[60] Talha MA, Goswami M, Giri BS, Sharma A, Rai BN, Singh RS. Bioremediation of Congo red dye in immobilized batch and continuous packed bed bioreactor by Brevibacillus parabrevis using coconut shell bio-char. Bioresource Technology. 2018;**252**:37-43

[61] Geed SR, Kureel MK, Giri BS, Singh RS, Rai BN. Performance evaluation of malathion biodegradation in batch and continuous packed bed bioreactor

(PBBR). Bioresource Technology. 2017;**227**:56-65

[62] Kapdan IK, Kargi F, McMullan G, Marchant R. Biological decolorization of textile dyestuff by Coriolus versicolor in a packed column reactor. Environmental Technology. 2000;**21**(2):231-236

[63] Yum KJ, Peirce JJ. Biodegradation kinetics of chlorophenols in immobilized-cell reactors using a white-rot fungus on wood chips. Water Environment Research. 1998;**70**(2):205-213

[64] Torres-Farradá G, Manzano AM, Ramos-Leal M, Domínguez O, Sánchez MI, Vangronsveld J, Guerra G. Biodegradation and detoxification of dyes and industrial effluents by Ganoderma weberianum B-18 immobilized in a lab-scale packed-bed bioreactor. Bioremediation Journal. 2018;**22**:1(1-2):20-27

[65] Losi ME, Giblin T, Hosangadi V, Frankenberger WT Jr. Bioremediation of perchlorate-contaminated groundwater using a packed bed biological reactor. Bioremediation Journal. 2002;**6**(2):97-103

[66] Banerjee A, Ghoshal AK. Biodegradation of an actual petroleum wastewater in a packed bed reactor by an immobilized biomass of *Bacillus cereus*. Journal of Environmental Chemical Engineering. 2017;**5**(2):1696-1702

[67] Kureel MK, Geed SR, Giri BS, Rai BN, Singh RS. Biodegradation and kinetic study of benzene in bioreactor packed with PUF and alginate beads and immobilized with bacillus sp. M3. Bioresource Technology. 2017;**242**:92-100

[68] Nikakhtari H, Hill GA. Hydrodynamic and oxygen mass transfer in an external loop airlift bioreactor with a packed bed. Biochemical Engineering Journal. 2005;**27**(2):138-145

[69] Patel BP, Kumar A. Biodegradation of 4-chlorophenol in an airlift inner loop bioreactor with mixed consortium: Effect of HRT, loading rate and biogenic substrate. 3 Biotech. 2016;**6**(2):117

[70] Bertollo FB, Lopes GC, Silva EL. Phenol biodegradation by *Pseudomonas putida* in an airlift reactor: Assessment of kinetic, hydrodynamic, and mass transfer parameters. Water, Air, & Soil Pollution. 2017;**228**(10):398

[71] Azeez TO, Onukwuli OD, Araromi DO, Arinkoola AO, Salam KK, Iwuji SC, et al. Optimization of bioremediation of cheese whey with the activity of Klebsiella pneumonia using response surface methodology. International Journal for Science and Engineering Technologies with Latest Trends. 2013;**11**(1):10-21

[72] Meng F, Chae SR, Drews A, Kraume M, Shin HS, Yang F. Recent advances in membrane bioreactors (MBRs): Membrane fouling and membrane material. Water Research. 2009;**43**(6):1489-1512

[73] Visvanathan C, Thu LN, Jegatheesan V, Anotai J. Biodegradation of pentachlorophenol in a membrane bioreactor. Desalination. 2005;**183**(1-3):455-464

Chapter 10

Phytoremediation of Effluents Contaminated with Heavy Metals by Floating Aquatic Macrophytes Species

Cleide Barbieri de Souza and Gabriel Rodrigues Silva

Abstract

The progress of urbanization and technologies led to the rise of anthropogenic activities, which consequently have high production of pollutants, affecting ecosystems, including aquatic biomes. One of the contaminating forms that cause environmental impact is heavy metals, which are produced in large quantities by inappropriate disposal of batteries, residential, industrial, agricultural and mining waste. Such components generate bioaccumulative effects, classifying them as dangerous elements that must be removed from environment. However, in species such as plants, this bioaccumulative effect can be exploited, aiming a biotechnological and bioengineering application to remove metals, called phytoremediation, employing floating aquatic macrophytes, which have high potential due to their properties retaining contaminants. Results obtained were conclusive for adaptation of *Eichhornia crassipes* and *Salvinia auriculata* as better phytoremediation agents, respectively, while *Lemna minor* and *Pistia stratiotes* fit better in biomonitoring, which have resistance to certain concentrations of metal when related to Cd, Hg, Zn, Ni and Pb.

Keywords: environmental bioengineering, biotechnology, phytoremediation, water, heavy metals, floating aquatic macrophytes

1. Introduction

In contemporary society, the great amount of harmful elements produced that come to nature has contributed greatly to environmental degradation, such aggressions to the environment are due to several factors, mainly by the pollutants and substances that affect the planet and its spheres (atmosphere, lithosphere, hydrosphere), as well as affect the biosphere, which participates in these different levels and aggregates all life and its different ecological niches. Therefore, it is necessary to find alternatives of reduction and even exclusion of these pollutants that affect our planet in a devastating way [1, 2].

Since the harmful damages to the environment from human activities were discovered, we look for sources and methods that can maintain sustainability, thinking about ecologically sustainable projects, making biodegradable products, looking for options to reduce pollutants and contaminants, other substitutions for extraction of material resources aimed at less damaging the ecosystem, creating environmental remediation mechanisms, among many other extraordinary ideas that have been carefully studied and applied successfully, but which in most cases can not completely reverse what has already been destroyed [1].

Many techniques have been applied obeying the principle of sustainability, achieved through the use of the reasoning of distinct areas of knowledge, which, through the same purpose, manage to create products, methods and innovations that lead to some environmental benefit. An area of extreme importance that contributes greatly to the development of this type of clean technology so necessary for the environment is Biotechnology, in which the knowledge of biological systems, organisms and their products are used for the insertion in processes that can generate high impact in the society [3], including by implementing biotechnological processes that can reduce contaminants or even improve environmental issues on a large scale.

When dealing with techniques that aim at direct environmental improvement, bioremediation is one of the most effective biotechnological alternatives to combat pollutants, especially when dealing with larger cases, such as in the case of an oil spill. Such technique consists in the removal or transformation of toxic elements by living organisms that when associated with a medium confer some characteristic that holds the aggressor, making it impossible for this element in its harmful form to still be in dispersion or contact with other biological and natural resources [4, 5].

"Brazil has a privileged situation in relation to water availability (...) the idea of abundance served for a long time as a support to the culture of wasted water available, and then acquired low value as a resource (...)" [6], due to the excessive water loss not only from Brazil, but worldwide, the minimum measures that can be implemented are those of environmental remediation, following the sustainable reasoning.

Plants used in the phytoremediation technique can be used in the terrestrial environment to remove organic or inorganic pollutants such as: toxic metals, chlorinated solvents, petroleum hydrocarbons, polychlorinated biphenyls and even radionuclides [7], as well as several species are used in the aquatic environment [8], where one of the genera that can be manipulated are the aquatic macrophytes, which has presented great expectations related to the remediation, providing numerous benefits in the extraction mainly when they deal with its bioaccumulative principles of heavy metals [9] in the media contaminated, which will be more exploited in this work, since they have functions that can be used for the restoration of water resources.

In this context, the objective of this work was a review for the evaluation of phytoremediating properties of aquatic macrophytes against heavy metals, such as: Cd, Hg, Zn, Ni and Pb according to the bioaccumulative capacity of four species of macrophytes. These metals were chosen because of their toxic properties or their ability to contaminate the environment, and they are present among pollutants with higher indices in wastewater and high metal toxicity [8, 9], while the selected plants were the most prominent in scientific articles.

2. Methodology

This research was carried out through a collection of information with the comparative function of studying and discussing the potential of phytoremediation of some species of floating aquatic macrophytes, made through the cross - linking of data from scientific articles from sources such as PubMed and Scielo, academic papers of high relevance and journals of impact, in order to cover in a global way the state of the art of this study.

3. Hydrical contaminants and heavy metals

The intense urbanization and increase of anthropogenic activities, coupled with the surplus population growth, have contributed to the scarcity of primary natural resources, extinction of species, among innumerable damages caused by urban and industrial pollution that can be cited, such as the increase in the incidence of diseases of the nervous system, often caused by heavy metals from vehicles, mining, metallurgy and electroplating; the increase in particulate matter related to the increase in the rate of respiratory diseases, besides neoplasias that may arise from several origins, among these there are already hypotheses of some harmful pollutants and others already proven that are still used in our daily life [1], as an example the nickel, an element in great abundance used in metal alloys, electroplating, composition of batteries among other products, being related to episodes of contact dermatitis and indications of carcinogenicity according to the increase of its concentration [9].

Heavy metals are inorganic compounds, some even considered as essential micronutrients for the metabolism of plants and other living beings, but when in high concentrations (varying according to the referred compound) can generate toxic effects on organisms, it is important to note that these effects may vary according to circumstances such as the organism tolerance, pH, presence of other ions that interfere with bioavailability, among other issues that may interfere with the result of contact with the element [10, 11]. With this toxicity, adverse effects are caused in plant physiology and metabolism, interacting mainly with their enzymes; since the plants are the organisms that theoretically have direct contact with these elements, thus they carry forward the damages of the bioaccumulation. These compounds can be captured by plants when they come in contact with soil or water that is contaminated with waste, emphasizing the industrial and agricultural [12], the latter cited is widely used to exemplify the accumulation of these materials in the food chain. The response of plants to contact with heavy metals depends on the concentration and exposure to them, presenting some phytotoxicity traits as reduced growth, especially the root system is more affected, chlorosis and leaf necrosis followed by traces of senescence and abscission, which changes lead to lower nutrient uptake and interfere with the biomass acquired [13].

4. Environmental biotechnology and bioengineering

Biotechnology is situated, due to its range of possibilities, among one of the main tendencies to the vision of the future, always with a look ahead of the possible methods to improve the quality of life, besides using resources in a less aggressive way. Aiming at the progression of mankind and at the same time environmental preservation, several biotechnological routes were created, as examples are much cited fields of food production, energy generation, prevention of environmental pollution and remediation areas [4].

Bioengineering is widely used in the environmental aspect aiming at the maintenance of resources, commonly present in soil erosion recovery processes along riverbanks [14], but can also be applied to water bodies, aiming at reestablishment with interventions designed from problems such as discharge of domestic, industrial and agricultural effluents and diffuse pollution, making possible an improvement in the water body in question and consequently of the biota that inhabits it [15].

Among these processes were created "phytotechnologies," alternative methods of remediation with plant utilization, aiming at the cleaning or stabilization of contaminants, this being an important function, which becomes more effective when implemented with other mechanisms and primary remediation techniques, promoting their decontamination capacities together [16].

There are also different techniques associated with measures to promote phytoremediative capacities that have promising results, such as the use of rhizobacteria associated with plants and transgenic practices adopted to overcome their absorptive or resistance to toxicity [17]. The application of phytoremediation is acceptable because it is a green technology, making phytoremediation a very useful technique in environment-related biotechnology, combining the characteristics of plant species for gradual decontamination.

5. Vegetable resistance to metals

Plants to acquire resistance to heavy metals have several mechanisms, often varying the tolerant characteristics of the plant to different metals, in order to prevent these harmful components from reaching their tissues and cells, interacting in metabolic processes and leading to signals toxicity, although not completely elucidated, some part of the mechanisms is known. Among the effects that counteract the damages of heavy metals are mentioned: in the *E. camaldulensis* species when exposed to 45 $\mu mol/L^{-1}$ cadmium there is increase of carotenoids (related to the tolerance to oxidative stress), and there is also an increase in the thickness of the epidermis and root endoderm according to the increased doses of the metal and the decrease in the thickness of the mesophyll and leaf limb related to the decrease of the photosynthetic capacity [7].

Oxidative stress is the toxic effects generated in plants through free radicals or reactive oxygen species (ROS) such as hydroxyls, superoxide anions, hydrogen peroxides [18] and may cause changes in proteins, DNA and membrane lipid peroxidation, which will interfere in the chlorophyll production inhibiting the production of α-aminolevulinic acid, in addition to demonstrating that the plants exposed to cadmium had lower water potential than the control group [7].

Reactive oxygen species are constantly produced in plants, this condition can be stimulated when in contact with a situation of homeostatic imbalance, as is the case of the presence of heavy metals, however the vegetables have an antioxidant defense system, to reduce the harmful effects the cells are among the antioxidant metabolites: ascorbate, glutathione, β-carotene, α-tocopherol that sequester or promote the degradation of ROS [19]. One of the main mechanisms related to this resistance being studied are phytochelatins.

5.1 Phytochelatins

Also called PCs (Phytochelatins), polypeptides are composed, most often by glutamic acid, glycine, and with a large amount of cysteine, and the presence of other amino acids may occur. Synthesized from glutathione (GSH) in steps catalyzed by the enzyme PC synthase, the increase of PC synthesis occurs by the contact of the plant with metal ions, inducing the enzyme PC synthase to exert greater activity [20].

There is greater involvement of research involving Cu, Zn, Ni, As and especially when it comes to cadmium, there is an increased expression related to PC, both in angiosperms, gymnosperms and pteridophytes and even bryophytes, PC are the main products associated with resistance because they act as metallic chelants, so the plants that have PCs in greater quantity may acquire greater resistance to the contaminated environment. On the other hand the functions of phytochelatins in the absence of heavy metals are not elucidated, although there are hypotheses of the participation of PCs in processes metabolic enzymes that lack metals [21]. Other non-plant organisms have PC biosynthesis, such as yeasts, algae and nematodes [22].

It is worth emphasizing that the metals that can be complexed by the phytochelatins are not all, being more effectively chelating Cadmium (by the incorporation of sulfides and formation of a cadmium sulfide coated with phytochelatins), Pb, Ag and Hg being the last three verified the formation of complexes with phytochelatins *in vitro*; Cd and Cu showed the formation of complexes with PC *in vivo*. The mechanism of metal tolerance attributed to PCs is due, in the presence of the metal, to the induction of the enzyme Phytochelatin synthase, which uses glutathione as a substrate for PC formation, after which the ion is transported and isolated in vacuoles, where they form complexes together to sulfides or organic acids [21, 22]. In addition, like the PC's, the amino acid cysteine that contains the sulfur in its structure is present in other clusters also called metallothioneins, which have in common a sequence of tripeptides, two of which are cysteine and a variant, having their binding capacity due to the sulfhydryl group present, with affinity for several metallic ions such as Zn, Hg, Cd, Cu, among others, for transport, storage and protection functions against metals and free radicals [23].

5.2 Distinct mechanisms of metal resistance in plants

There are some mechanisms different from the PCs that make the plant acquire more tolerance, as it is the case of the increase in the thickness of the wall root tissue, being the endoderm and exoderme root one of the main tissues to retain metals, association with microorganisms like fungi in the root region (mycorrhizae), can also secrete root exudates, compounds such as organic acids that complex the metal cations reducing the extracellular activity of these, are capable of suppressing the transport system, thus reducing its absorption, heat shock proteins (HSP) that have already been associated with protection of the plasma membrane, complexation or chelation by ligands pointed out in researches such as phytochelatin, metallothionein and organic acids that reduce the activity of the metal in the cytosol, which when ligated with the same are transported to vacuoles and stored, all these are mechanisms of immobilization, exclusion, repair of damages or complex, and have not been well elucidated yet, so they may have other relations and functions between them [18]. Due to these attributes a plant can become a good species for the phytoremediation technique.

6. Phytoremediation

Plants are organisms that maintain a good part of the ecosystem due to their beneficial properties to the means in which they are: phytodegradation, phytostimulation, volatilization, phytoprocessing, rhizofiltration, and also serve as a base to cover the area of environmental remediation, called phytoremediation. Their properties range from natural bioindicators to organisms that can be applied to a bioremediative technique; having several functions to remove contaminants, for these reasons, it has been proven that studies are still needed to identify species capable of acting more effectively [8, 24].

Among the main elements that can be remedied by plants are petroleum hydrocarbons, agrochemicals, explosives, chlorinated solvents and industrial byproducts, organic or even inorganic substances [25]. The focus of this chapter is related to heavy metals, which fit the last category mentioned and can also be related to the activities of mining [26], causing long-term toxic effects to those who eventually come into contact with them.

Phytoremediation mechanisms are widely used in the environment contaminated by principles of bioaccumulation, occurring commonly in soil, however, are not only limited to this site, due to the variety of species in the flora, such as the extraction of pollutants from the aquatic environment by aquatic macrophytes, which according to the definition of phytoremediative capacities, those that have greater resistance to the toxicity of the environment, have great capacity of biomass generation in a shorter interval of time, becoming great options for gradual decontamination of the environment, having substantial action in the case of effluents containing metals [8, 27].

The phytoremediative features that give the plant favorable conditions for its use, according to [8] are: high growth, high biomass production, vigor and resistance to toxicity, in which the macrophytes fit perfectly. Of the many benefits that can be cited from the use of plants included in these aquatic macrophytes, [28] states that "Remedial potential in various environments may under certain circumstances be a good alternative to habitat decontamination methods" the possibility of exploring the theme in a way that demonstrates the important collaborations of this application for the survival of a water ecosystem and its benefits to a contaminated environment, mainly exalting the relation of its efficiency in the extraction of heavy metals, always evaluating its advantages of removal, efficacy, selectivity and cost of the process.

Thus the phytoremediative process is due to an excellent biotechnological technique capable of overcoming in terms of expenses and sustainability, since it aims at improvements with the minimum impact to nature [27].

6.1 Aquatic macrophytes

Among the aquatic plants are the macrophytes, their group encompasses several aquatic plants that are generally divided by their classification of life form, being categorized in: free floating, submerged rooted, rooted with floating and emerging leaves [29]. It also shows that in the neotropical countries, Brazil is among one of the largest producers of scientific content related to macrophytes, solidifying the interest of the use and knowledge of these species that has great ecological importance. Studies have shown that the richness of the species of floating aquatic macrophytes comes from the morphoclimatic characteristics, through the relation between the hydrological characteristics and the diversity patterns of these plants [30]. The regions where they have predominance are, respectively, wetlands (wastewater treatment areas), lakes-ponds, and also river-streams; there are other ecosystems where they are found as the case of the marine environment, however they do not present population as expressive as the first three cited previously [29].

Some macrophytes may be considered "weeds," due to their resistance and exponential vegetative growth characteristics, which on the one hand become important factors that affect the biomass control of the species treated as pests. The *Eichhornia, Salvinia,* and *Pistia* genera, when exposed to pesticides that are often effective to other plants, do not demonstrate such toxicity in some types of superweeds, these resistance attributes that make them good organisms for the environmental remediation technique [31].

6.2 Free floating aquatic macrophytes

Macrophytes may be associated with phosphorus and nitrogen concentrations in water, and their development is generally favored by high temperatures. When they are classified as floating they have characteristics such as high primary productivity, commonly found in waters with moderate current velocity, and are limited by the oscillation of the water level, since this can lead to death of the plant population of these species. Within this classification of floating macrophytes are some species, such as: *Eichhornia crassipes, Pistia stratiotes*, *Salvinia auriculata* and *Lemna minor* [32].

6.2.1 Eichhornia crassipes

The species *E. crassipes*, known as "aguapé" or "water hyacinth," is a floating aquatic macrophyte, native to South America and belongs to the family Pontederiaceae [31], highly tolerant to polluted environments and effective for extracting nutrients and metals such as Cd, Pb and Hg. Research has shown that for elements such as Hg, it has great absorption properties, with only the characteristic of chlorosis in the plant when exposed, without drastically affecting its growth over a long period with a certain quantity of mercury, counting on this attribute even in the remediation of water with high levels of K and NO^{3-} [33]. Also, it was demonstrated the ability of this species to absorb many other elements such as Cd, Cu, Ni, Pb, Zn, with copper being the most absorbed in the root region, presenting a high value bioconcentration factor; for lead, when presented at low concentrations demonstrates large accumulation at roots, not finding values as high for the other metals in this study [34]. In another comparative study between cadmium values absorbed by Aguapé and Salvinia, the Aguapé was more tolerant to the metal, and the author associated this resistance with the higher production of thiol compounds by *Eichhornia crassipes* [35]. In another comparative research between *E. crassipes*, *P. stratiotes* and *S. molesta*, Aguapé has an advantage over the characteristics of phytoremediation of effluents due to its large biomass production capacity and the accumulation of several elements at the same time, increasing its biomass by 4.5× in only 30 days in the quoted experiment [36, 37].

The absorption capacity of Zn(II) and Cd(II) measured in the biomass (stem and leaf) of this macrophyte is respectively 9.3 and 12.4 $mg.g^{-1}$, as maximum capacity of metal in the plant [9]. As results of metal absorption this species succeeded in the bioaccumulation of the metals evidencing greater percentage of removal respectively of Zn, Pb and Cu [38].

6.2.2 Pistia stratiotes

The species *P. stratiotes*, commonly known as "water lettuce", considered as a pest because of its resistance and replication capacity, even by the characteristic of regrowth [31]. It belongs to the Araceae family, monocot, commonly distributed in tropical and subtropical countries, it is demonstrated that it is a good accumulator of Copper (Cu), although other elements can also be captured by it [39].

Studies have demonstrated bioaccumulation values at distinct points of the Guarapiranga reservoir localized in SP/Brazil, for different macrophytes. The values of bioaccumulation between two macrophytes in three different sites of the same reservoir were evaluated, being high values in the *E. crassipes* of the As, Co, Cu, Mn, Ni elements, as well as the accumulation of this species, including the element Zn, which has significant difference between the values of *Pistia stratiotes*. Although Pistia, at different collection sites, it was able to obtain results similar to or even slightly larger than that of the Aguapé for Cu, Pb and Ni elements [40]. The accumulation value for Zinc is also highlighted in another study in which at the end of the third stage of the experiment Zn values were almost zero from an initial solution with 0.01 $mg.L^{-1}$, indicating that it may also be employed in remediation [41].

6.2.3 Salvinia auriculata

This species of macrophyte in the family Salviniaceae is a plant in which each individual has three hairy verticillate leaves, being two floating, has high capacity of absorption of many elements as the other plants, however studies have shown the response of potential bioindicator of this species to cadmium, because even at a solution concentration of 10 μmol.L^{-1}, it presents very great toxicity, consequently leading to oxidative stress, foliar necrosis and death. These results state that the possibility of *Salvinia auriculata* remediation of Cd is not as effective, leading to a possible ecological indicator for this species, since the accumulation data presented are not high due to its sensitivity when exposed to metal [42]. In another study it is also proposed that when the amounts of zinc in the water are above the amount allowed by the legislation, 5 mg.L^{-1} this species already shows signs of toxicity, a result that is biased toward its use in biomonitoring (reaction bioindicator) of contaminated sites, whereas for Pb it is considered to be a good accumulator mainly in the submerged part of the plant. This absorption occurs through the plasma membrane and presented a better accumulated value in the concentration of 5 mg.L^{-1} in the solution, and although it decreased its biomass, their clonal reproduction was not completely lost [43]. In a comparative study of this species with two more of the same family, *S. molesta* and *S. natans*, the results obtained indicate that the bioaccumulation values of *S. auriculata* are not the largest compared to these two species for the metals Cd, Hg, Pb, Ni and Zn, often meeting as intermediary between them, although their employability in phytoremediation is still possible [44].

6.2.4 *Lemna minor*

Also known as "duckweed", has a cumulative capacity, in a study carried out by [45], makes a comparison with the *Azolla filiculoides*, obtaining results of good Mn accumulation by *Lemna*, although higher concentration of other metals in *Azolla*, and concluded that both plants are candidates for phytoremediation of contaminated and wastewater. Another study demonstrated that in addition to wastewater treatment the macrophyte can be used for aquatic toxicity tests, taking into account its sensitivity to heavy metals and herbicides, as well as the possibility of being used as animal feed (pigs, poultry, cattle, among others) provided that do not contains any toxic substances included [46].

In an experiment done by [47] the efficiency of Hg removal is tested by comparing the species *E. crassipes* and *Lemna minor*, proving a higher uptake by the first, both with indications of morphological changes due to toxicity, and although the rapid reproduction of *Lemna minor*, the one with the most resistance is *E. crassipes*. According to [48], when compared to other plant species *Lemna minor* is not the first desirable one to be chosen, since its performance does not support peaks of contaminant concentration and grows well in places with low concentration pollution although these are also of risk because its bioaccumulation is great, it would be ideal to use it for attributions such as polishing the water after pretreatment or even using its sensitivity characteristics for biomonitoring. Another comparison is made between *P. stratiotes*, *L. minor* and *Spirodela intermedia*, where they obtained excellent absorption values, especially *P. stratiotes*, although the *L. minor* species did not survive at the end of the experiment, the theory proposed by the author due to vegetal cover in the water layer that prevents the increase of oxygen dissolved in the recipient, leading to the death of individuals [49]. Research indicates that there is the capacity of using this plant also to increase the oxygenation of deep waters, called phytodepuration [50].

6.3 Phytoremediation mechanism

The remedial capacity of a plant is directly linked to its absorptive and cumulative potential of metals, with the help of proteins from the plants themselves acting as chelators, thus being able to isolate these substances in vacuoles with the intention of reducing toxicity to the organism itself [12]. Through phytochelatin that acts as chelator and also of other proteins, the metals join to this protein and form complexes, being able to be transported to vacuoles, as a form of defense for the compound does not attack the vegetable organism [51].

According to the remediation mechanisms provided by the plants, the characteristics of phytoextraction, phytodegradation, phytostabilization, phytostimulation, phytovolatilization, phytotransformation and rhizofiltration are included [52].

6.3.1 Phytoextraction

Absorption of contaminant, with transport and storage at a particular plant site (roots, leaves, stems), it is also indicated that heavy metals hyperaccumulating plants are commonly used in this technique, and phytoextraction becomes effective only if the plants are removed before and during the degradation process, soils contaminated by Ag, Cd, Co, Cr, Cu, Hg, Mn, Mo, Ni, Pb, Zn and the radionuclides of Sr., Cs, Pu and U (unstable chemical elements that emit radiation) can be remedied by this technique [53, 54].

6.3.2 Phytodegradation

Due to the mechanisms of the plant, some products are degraded by the metabolism processes and are usually organic, but there are some inorganic compounds that can also be degraded, among them: TNT (Trinitrotoluene), DDT (DichloroDiphenylTricloroethane), TCE (Trichlorethylene), HCB (Hexachlorobenzene) and PCB's (Polychlorinated Biphenyls) [54].

6.3.3 Phytostabilization

When it is not possible the degradation of the contaminating element, precipitation of this component occurs in the rhizospheric region, between the root and the soil, applicable mainly when it comes to metals like Cd, Cr, Cu, Hg, Pb and Zn, so it is necessary tolerant plants for adequate stabilization [54].

6.3.4 Phytostimulation

The plant in the rhizosphere region produces substances that accelerate the degradation of the organic component, in addition to stimulating the proliferation and development of microorganisms such as bacteria and fungi, which may help in the process of uptake and degradation of contaminants, one of the main compounds remedied in these petroleum hydrocarbons [52].

6.3.5 Phytovolatilization

Process occurred to contaminants that may be volatilized after metabolization or biodegradation in the rhizosphere. It is also mentioned in a study that some compounds can be released by the direct action of the plant through the surface of the leaves, and that still they can be contaminants, being able to accumulate also in fruits and in the stem, among the inorganic components that can be volatilized are Se and Hg, although the use of this technique is still questionable for some authors [54].

6.3.6 Phytotransformation

The contaminant product is biotransformed, that is, converted to the less harmful substance, a feature that is very useful for the remediation of organic compounds, and this process can be understood as the metabolization of the pollutant in less toxic forms by the plant itself [52].

6.3.7 Rhizofiltration

When the phytoextraction is applied in a hydroponic medium, it has the name of rhizofiltration, it consists of the adsorption of the contaminants through the combination of the extraction and stabilization, is suitable for the removal of Pb, Cd, Cu, Ni, Zn, as well as some radionuclides [54].

6.4 Phytoremediative applications

As some of the many examples of decontamination made by plants, it can be demonstrated [55], where in a contaminated environment (Arroio de Santa Barbara, Rio Grande do Sul, Brazil) species of macrophytes of the families Asteraceae, Araliaceae, Poaceae, Araceae and Alismataceae were collected, including *Pistia stratiotes*. In this study, the concentration of bioconcentrated metals in these plants was measured. The obtained values proved that the remediation is effective because of its bioconcentration and resistance to the contaminants, and the metals studied and accumulated in this study were Zn, Cr, Pb, Ni and Mn.

A study by [56] suggested the technique of decontamination by plants and applied in wetlands with 20 L of effluent received by the Mussuré Creek, Paraíba, Brazil, to verify the efficacy of phytoremediation of three species. Divergent genus and species of plants have been studied, as *Lemna sp*., *Pistia stratiotes*, and *Eichhornia crassipes*, obtaining results such as color reduction, turbidity, biochemical oxygen demand (BOD), being efficient for treatment of domestic effluent, besides being a simple and low cost system.

Other contaminant removal applications can be described, such as those cited in [57], Indian mustard becomes effective in the removal of Cd, Cr, Cu, Ni, Pb and Zn; the sunflower that has accumulating property (in hydroponic means) of Pb, U and even Cs-137 and Sr-90 that are radioactive isotopes demonstrating a wide phytoremediative capacity of several different species.

In parallel, some authors cite different methodologies, through transgenic plants, such as the CYP2B6 gene insertion of the human cytochrome P450, inducing the tolerance of the rice plant, and when compared to a control group, conferred resistance to herbicides, reaching the result that the plant can effectively remove contaminants such as Metolachlor, a member of the chloroacetanilide herbicide family. This proposed editing of genes to extend resistant attributes of plants has been well discussed in articles for biotechnological purposes [58].

6.5 Phytoremediation advantages by floating aquatic macrophytes

Phytoremediation is of great value when compared to other methods of environmental decontamination because its variety of contaminants that can be captured from the environment is very broad, including metals, radionuclides, various categories of organic pollutants, functioning as an efficient ecologically acceptable method with very low cost, although the remediation characteristics are dependent on the concentration of pollutants that generate toxicity to the plants that will come into contact, logically varying species, type of pollutant and environmental characteristics, so it is necessary in these cases to use other methods of direct decontamination [59].

This method is among one of the biotechnological techniques available for environmental decontamination, although it is more used in tropical climate countries, its application in Brazil also has a wide expectation due to the vast flora and temperature suitable for the development of the plant specimens, which consequently facilitates success when adopting the phytoremediation technique. As a mean of mitigating the situation of water bodies contamination, there are already studies proving the efficacy and relevance of the phytoremediative process of floating aquatic macrophytes, their applications if well studied should lead to a gradual decontamination of the environment, as long as their characteristics are analyzed before to apply them *in situ*. According to the studied macrophytes, one of the most outstanding in the literature was the *Eichhornia crassipes* species, which has resistance attributes, including to pesticides [31], their replication rate and accumulation are favorable for the metals analyzed [38, 40], however it is necessary to emphasize that the other species may still be appropriate for the environmental remediation functions, since it is noted that the values of bioaccumulation, absorption and tolerance to metals have divergences in each experiment. Such plants are effective in the capture of heavy metals in the aquatic environment, and may even serve as biosensors, such as the species *S. auriculata* [41] and *L. minor* [46, 48] that are more suitable for this task or for the polishing of the water, according to the morphological characteristics presented when in contact with potential contaminants. The four species (*Eichhornia crassipes, Pistia stratiotes, Salvinia auriculata* and *Lemna minor*) are of great value in the aspect of contaminant removal [38, 41, 44, 45] because they facilitate the post-capture process of metals, where floating macrophytes take advantage over the rooted ones and contribute to the process of plant removal, consequently the extraction of the metals from the medium is facilitated.

There are some difficulties encountered in the application of the technique such as: time to obtain the results not immediate, doubt about the correct techniques of plant fate post treatment still existing and not be completely profitable for patenting as has been shown by [7], on the other hand, there are advantages of this method like the public acceptance and its attractiveness, the possibility of being applied in preventive methods, or even allied with other types of remediation, such as its excellent polishing capacity in already functioning water treatment systems, besides its economical characteristic satisfactory.

Although it is considered to be a green and ecologically acceptable technology, special care should be taken, such as the need for research and technological development, to verify pre and post-harvest phases of biomass from the site being treated, such as: the best species to remediate certain contaminants, to be aware of the route and transformation of the contaminants in the body of the vegetable, to destine to the biomass in suitable places. All these requirements are required for detailed strategies to achieve results as expected to avoid transferring contamination to other receptor bodies [7]. Many methods are used for decontamination; however, the use of vegetable mass for such a function has a great advantage over costs.

7. Destinations of plant biomass after treatment

After remediation of the contaminated effluent or body of water, the biomass incorporates the metals and retains them, that is, it still has the bioaccumulated heavy metals that need to be removed from the contact with the medium and can be submitted to manual processes or even mechanics for the withdrawal of the vegetal specimens, reaching the stage of isolation of the vegetal mass, thus avoiding a new dispersion of the contaminants post extraction. However a final disposal is required, often subjecting it to harmful environments such as the deposit of these contaminants in landfill and incineration; the latter technique is mentioned as a proposal to reduce the volume of plants and to generate energy by means of heat, also contributing to the removal of metals [8], However, research suggests the incorporation of the biomass used in the treatment of ceramic blocks, proposing an efficient method of transformation of the products obtained, promoting a new environmentally acceptable use, without generating more contaminants and dispensing with incineration

[26], also studies affirm that the incorporation of the vegetal components in the ceramic sector is possible because the clay favors in the inertization of the as a way of preserving natural resources and avoiding further degradation of the environment. In the same work are mentioned researches of incorporation of macrophytes used after the technique of phytoremediation with large amounts of copper and chromium, proving in both the viability of inclusion of a percentage of the biomass in the clay leading to the consequent inertization of metals [60].

Measures such as the use of biomass in boilers, furnaces and for electric power generation, as well as the production of biogas in biodigesters, are proposed to avoid the final disposal of these macrophytes as solid residue, however, when used after the phytoremediation method, all caution is required to remove toxic components and avoid recontamination [60]. Although the phytoremediative method is a great technique and low cost, it has some contradictory factors that cause doubt in the application, one of them is the relation of the correct disposition of the contaminated material.

The elaboration of this work stems from the need to improve the study of phytoremediative capacities of certain species of aquatic macrophytes, especially those of the floating type (not rooted), due to the benefits brought to their processes of decontamination or removal of heavy metals, this is a subject which can be more explored and there is a certain lack in the technical-scientific scope related to this topic. With the present work, the characteristics of these macrophytes can be praised with in-depth study, relating their respective adaptations that lead to methods that may be less aggressive to treat water contaminated with metallic ions of the environment. The choice of the theme is also related to the diversity of the flora of aquatic macrophytes in the Brazilian region and the ideal temperature for growth, cited by several authors [30, 61, 62], among many other researchers who reaffirm the vast flora existing in Brazil, which including, is being increasingly investigated for phytoremediative aspects, due to the conditions of the numerous pollutants generated. Since the final disposal is considered one of the crucial procedures for phytoremediation, and seen that if not applied correctly can lead to a rebound contamination, there is a need for emphasize final destination processes in forward researches, suggesting an improvement of the phytoremediative technique if the appropriate destination is taken, reaching in that way the original idea of decontaminating phytotechnologies, amplifying percentages of removal [60].

8. Conclusions

Among the four species of floating aquatic macrophytes with phytoremediation potential cited in this study, *Eichhornia crassipes* stands out due to the number of studies and data available and a great capacity of biosorption and mass production combined with resistance to contaminants, although this species has great values of accumulation in relation to the others, it is unlikely to affirm that it would be the best alternative in all situations, questions of environmental temperature, pH, bioavailability, and mainly time of extraction related to metals and their interactions, biological oscillations that cause variations between each scientific experiment and that should not be disregarded, require more study to adapt more concrete values of absorption to each species. The potential uses of water hyacinth have been cited in studies such as their ability to remove radionuclides, wastewater from a wide range of sources (industrial, petrochemical, metallurgic, agriculture, pharmaceutical and personal care products waste, herbicides, even Polycyclic Aromatic Hydrocarbons can be included, recognized by your higher efficiency for being a hyperaccumulator, that can reduce turbidity by 92,5%, Cd by 97,5%, Ni 95,1% and Hg 99,9%, results acquired in researches depending of the concentrations of metal in water [63].

Acknowledgements

The authors would like to thank the Lusíada University Center (UNILUS) and the Nucleus of study and research in biotechnology and molecular biology (NAPBBM).

Conflict of interest

The authors, Cleide Barbieri de Souza and Gabriel Rodrigues Silva, declare that there is no conflict of interest that might constitute an impediment to the publication of this research work.

Notes/thanks/other declarations

Nothing to declare.

Appendices and nomenclature

Ag	silver
As	arsenic
BOD	biochemical oxygen demand
Cd	cadmium
Co	cobalt
Cr	chromium
Cs	cesium
Cu	copper
DDT	dichlorodiphenyltrichloroethane
DNA	deoxyribonucleic acid
GSH	glutathione
HCB	hexachlorobenzene
Hg	mercury
HSP	heat shock proteins
K	potassium
Mn	manganese
Mo	molybdenum
Ni	nickel
NO_3	nitrate
Pb	lead
PC	phytochelatin
PCB	biphenyl polychlorinated
Pu	plutonium
ROS	reactive oxygen species
Se	selenium
Sr	strontium
TCE	trichlorethylene
TNT	trinitrotoluene
U	uranium
Zn	zinc

Author details

Cleide Barbieri de Souza* and Gabriel Rodrigues Silva
Lusíada University Center (UNILUS), Santos, São Paulo, Brazil

*Address all correspondence to: cleidebarbieri@gmail.com

References

[1] Brilhante OM, Caldas LQA. Gestão e avaliação de risco em saúde ambiental. Rio de Janeiro: FIOCRUZ; 1999. 155 p

[2] Derisio JC. Introdução ao Controle da Poluição Ambiental. 4th ed. São Paulo: Oficina de Textos Editora; 2007. 224 p

[3] MMA. Convenção de Diversidade Biológica. artigo n°02. Available from: http://www.mma.gov.br/estruturas/sbf_chm_rbbio/_arquivos/cdbport_72.pdf

[4] Schenberg ACG. Biotecnologia e desenvolvimento sustentável. Estudos Avançados. 2010. FapUNIFESP (SciELO):07-17. DOI: 10.1590/s0103-40142010000300002

[5] Pakshirajan K, Rene ER, Ramesh A. Biotechnology in environmental monitoring and pollution abatement. BioMed Research International, [s.l.]. 2014;**2014**:1-4. DOI: 10.1155/2014/235472

[6] Setti A et al. Introdução ao gerenciamento de recursos hídricos. 2nd ed. Brasília-Distrito Federal: ANEEL. ANA; 2001

[7] Marques M et al. Desafios técnicos e barreiras sociais, econômicas e regulatórias na fitorremediação de solos contaminados. Revista Brasileira de Ciência do Solo [en linea]. 2011;**35**:1-11

[8] Lamego FP, Vidal RA. Fitorremediação: Plantas Como Agentes De Despoluição? Pesticidas: Revista de Ecotoxicologia e Meio Ambiente, [s.l.]. 2007;**17**:9-18. DOI: 10.5380/pes.v17i0.10662

[9] Lima JRA. Remoção De Metais Em Água Utilizando *Eichhornia crassipes* Na Forma *In Natura*, Biocarvão E Híbrido Magnético. Fevereiro, São Cristóvão: Universidade Federal de Sergipe; 2018

[10] Jarup L. Hazards of heavy metal contamination. British Medical Bulletin, [s.l.]. 2003;**68**(1):167-182. DOI: 10.1093/bmb/ldg032

[11] Uechi DA, Gabas SG, Lastoria G. Análise de metais pesados no Sistema Aquífero Bauru em Mato Grosso do Sul. Engenharia Sanitaria e Ambiental, [s.l.]. 2017;**22**(1):155-167. DOI: 10.1590/s1413-41522016142430

[12] Bücker-Neto L et al. Interactions between plant hormones and heavy metals responses. Genetics and Molecular Biology. 2017;**40**(11):373-386. DOI: 10.1590/1678-4685-gmb-2016-0087

[13] Brunetto G et al. Heavy metals in vineyards and orchard soils. Revista Brasileira de Fruticultura, Jaboticabal. 2017;**39**(2):e-263. DOI: 10.1590/0100-29452017263

[14] Sonego MF, Pinto MSN, Neto EG. Aplicação De Técnicas De Engenharia Natural Na Renaturalização De Rios Urbanos: Estudo De Caso No Rio Bacacheri. In: 16° CBGE (Congresso Brasileiro de geologia e engenharia ambiental). 1º ed. 2018. ISBN 978-85-7270-076-4

[15] Oliveira AG, Crispim MC. Water systems' revitalization: Interventions adopted in aquatic ecosystems in Brazil and worldwide. Gaia Scientia. 2017;**11**(1):297-306. DOI: 10.21707/gs.v11.n01a19

[16] EPA—Enviromental Protection Agency. Phytotecnologies for site cleanup. US, 2010. disponível em. Available form: https://clu-in.org/download/remed/phytotechnologies-factsheet.pdf [Accessed: Nov 20, 2018]

[17] Dowling DN, Doty SL. Improving phytoremediation through biotechnology. Current Opinion in

Biotechnology, [s.l.]. 2009;**20**(2): 204-206. DOI: 10.1016/j.copbio.2009.03.007

[18] Rodrigues ACD et al. Response mechanisms of plants to heavy metal pollution: Possibility of using macrophytes for remediation of contaminated aquatic environments. Revista Virtual de Química. 2016;**8**(1):272-276. DOI: 10.5935/1984-6835.20160017

[19] Barbosa MR et al. Geração e desintoxicação enzimática de espécies reativas de oxigênio em plantas. Ciência Rural, Santa Maria. 2014;**44**(3):453-460

[20] Melendez LB et al. Determinação de Metalotioneínas e Fitoquelatinas utilizando a técnica de HPLC-ICP-MS. Revista Virtual de Química. 2012;**4**(6):612-622. DOI: 10.5935/1984-6835.20120047

[21] Cobbett CS. Phytochelatins and their roles in heavy metal detoxification. Plant Physiology, [s.l.]. 2000;**123**(3):825-832. DOI: 10.1104/pp.123.3.825

[22] Inouhe M. Phytochelatins. Brazilian Journal of Plant Physiology. 2005;**17**(3):65-78. Available from: http://www.scielo.br/pdf/bjpp/v17n1/a06v17n1.pdf

[23] Arias ARL, Santos VG. Metalotioneína: Processos celulares e moleculares. Fiocruz. Rio De Janeiro: Cadernos Saúde Coletiva; 2008;**16**(4):701-716

[24] Borém A, Santos FR. Entendendo a biotecnologia. Viçosa: Suprema; 2008. 342 p

[25] Pires FR et al. Fitorremediação de solos contaminados com herbicidas. Planta Daninha, [s.l.]. 2003;**21**(2):335-341, ago. DOI: 10.1590/s0100-83582003000200020

[26] de Lima, Sonia Barbosa, de Souza, Débora Cristina, de Albuquerque, Felipe Pereira. Confection of ceramic bricks with biomass contaminated by cupper incorporation. Ambiência, [s.l.], v. 11, n. 1, p. 31-47, 2015. DOI: 10.5935/ambiencia.2015.01.02

[27] Rizzo PF et al. Remediation of feedlot effluents using aquatic plants. Revista de la Facultad de Ciencias Agrarias. Universidad Nacional de Cuyo, Mendoza. 2012;**44**(2):47-64. Available from: http://www.scielo.org.ar/scielo.php?script=sci_arttext&pid=S1853-86652012000200004&lng=es&nrm=iso

[28] Canhoto et al. Uso De Macrófitas Em Fitorremediação. 2010. Available from: https://estudogeral.sib.uc.pt/bitstream/10316/18836/1/Tese_Filipe%20Pratas.pdf [Accessed: Nov 20, 2018]

[29] Padial AA, Bini LM, Thomaz SM. The study of aquatic macrophytes in Neotropics: A scientometrical view of the main trends and gaps. Brazilian Journal of Biology, São Carlos. 2008;**68**(4 supl):1051-1059. DOI: 10.1590/S1519-69842008000500012

[30] Henry-Silva GG, Moura RST, de Oliveira Dantas LL. Richness and distribution of aquatic macrophytes in Brazilian semi-arid aquatic ecosystems. Acta Limnologica Brasiliensia, [s.l.]. 2010;**22**(02):147-156. DOI: 10.4322/actalb.02202004

[31] Martins D et al. Controle químico de *Pistia stratiotes*, *Eichhornia crassipes* e *Salvinia molesta* em caixas d'água. Planta Daninha. Sociedade Brasileira da Ciência das Plantas Daninhas. 2002;**20**(Sep):83-88. DOI: 10.1590/S0100-83582002000400010

[32] Thomaz S, Bini M. Ecologia e manejo de macrófitas aquáticas. Maringá, PR: EDUEM Editora da Universidade Estadual de; 2003. 342 p

[33] Caldelas C et al. Physiological responses of *Eichhornia crassipes* [Mart.] Solms to the combined

exposure to excesso nutrientes and Hg, Brazilian Journal of Plant Physiology. 2009;**21**(1):1-12

[34] Liao S, Chang N. Heavy metal phytoremediation by water hyacinth at constructed wetlands in Taiwan. Journal of Aquatic Plant Management. 2004;**42**:60-68. DOI: 10.12691/aees-1-5-4

[35] de Oliveira JA. Effects of Cadmium on Sulphur Uptake, Distribution and Assimilation in Water Hyacinth (*Eichhornia crassipes* (Mart.) Solms) and Salvinia (*Salvinia auriculata* Aubl.). D.S.: Universidade Federal de Viçosa; 1998

[36] Coelho JC. Macrófitas aquáticas flutuantes na remoção de elementos químicos de água residuária. Botucatu: Unesp; 2017. Available from: https://repositorio.unesp.br/bitstream/handle/11449/151881/coelho_jc_me_bot.pdf?sequence=3

[37] Dinardi AL, Formagi VM, Coneglian CMR, Brito NN, Sobrinho GD, Tonso S, et al. Fitorremediação. In: III Fórum de Estudos Contábeis. Limeira, SP: Centro Superior de Educação Tecnológica (CESET)—UNICAMP; 2003

[38] Luiz EP. Avaliação dos métodos de fitorremediação e biossorção na remoção de chumbo, cobre e zinco. Santo andré, SP: Ufabc. dissertação mestrado em ciencia e tecnologia ambiental; 2016

[39] Odjegba VJ, Fasidi IO. Accumulation of trace elements by *Pistia stratiotes*: Implications for phytoremediation. Ecotoxicology. 2004;**13**(7):637-646

[40] Coutinho SN et al. Estudo de bioacumulação de elementos potencialmente tóxicos em amostras de macrófitas aquáticas flutuantes do reservatório Guarapiranga, São Paulo. Rio de Janeiro: IPEN/CNEN/CETESB; 2017

[41] Cruz MB et al. Absorção De Metais Pesados Presentes Em Efluente De Mineração Por *Pistia stratiotes*. IX Congresso de Ecologia do Brasil. São Lourenço, MG; 13-17 September 2009

[42] Wolff G et al. The use of *Salvinia auriculata* as a bioindicator in aquatic ecosystems: Biomass and structure dependent on the cadmium concentration. Brazilian Journal of Biology, São Carlos. 2012;**72**(1):71-77. DOI: 10.1590/S1519-69842012000100009

[43] WOLFF G et al. Efeitos da toxicidade do zinco em folhas de *Salvinia auriculata* cultivadas em solução nutritiva. Planta Daninha. 2009;**27**(1):133-137

[44] Kumari S, Kumar B, Sheel R. Bioremediation of heavy metals by serious aquatic weed, *Salvinia*. International Journal of Current Microbiology and Applied Sciences, [s.l.]. 2016;**5**(9):355-368. DOI: 10.20546/ijcmas.2016.509.039

[45] Amare E et al. Field based investigation on phytoremediation potentials of *Lemna minor* and *Azolla filiculoides* in tropical, semiarid regions: Case of Ethiopia. International Journal of Phytoremediation, [s.l.]. 2017;**20**(10):965-972. DOI: 10.1080/15226514.2017.1365333

[46] del Pilar Arroyave M. La Lenteja De Agua (*Lemna Minor* L.): Una Planta Acuática Promisoria. Revista EIA Escuela de Ingeniería de Antioquia. 2004;**1**:33-38

[47] Jumbo MDCJ, Campoverde EDF. Fitorremediación mediante el uso de dos espécies vegetales *Lemna minor* (Lenteja de agua) y *Eichornia crassipes* (Jacinto de agua) em aguas residuales produto de la actividad minera. Cuenca: Universidad Politécnica Salesiana; 2012

[48] Obermeier M et al. The enzymatic and antioxidative stress response of *Lemna minor* to copper and a chloroacetamide herbicide.

Environmental Science and Pollution Research, [s.l.]. 2015;**22**(23): 18495-18507, 20 ago. DOI: 10.1007/s11356-015-5139-6

[49] Miretzky P, Saralegui A, Cirelli AF. Aquatic macrophytes potential for the simultaneous removal of heavy metals (Buenos Aires, Argentina). Chemosphere, [s.l.]. 2004;**57**(8):997-1005. DOI: 10.1016/j.chemosphere.2004.07.024

[50] Ferreira TCR, Ribeiro MC. Macrófitas aquáticas na remoção de microcistinas. MG, Brazil: Universidade Federal de Ouro Preto; 2009

[51] Gratao PL et al. Phytoremediation: Green technology for the clean up of toxic metals in the environment. Brazilian Journal of Plant Physiology. Mar. 2005;**17**(1):53-64. DOI: 10.1590/S1677-04202005000100005

[52] Tavares SRL. Fitorremediação em solo e água de áreas contaminadas por metais pesados provenientes da disposição de resíduos perigosos. 2009, 371 f. Tese (Doutorado em Ciências em Engenharia Civil)—Programa de Pós-graduação em Engenharia Civil. Rio de Janeiro: Universidade Federal do Rio de Janeiro/COPPE; 2009

[53] Oliveira D et al. Fitorremediação: estado da arte. Rio de Janeiro: CETEM/MCT; 2007. 49 p

[54] Mejía PVL et al. Metodologia para Seleção de Técnica de Fitorremediação em Áreas Contaminadas. Revista Brasileira de Ciências Ambientais. 2014;**31**:97-104

[55] Demarco CF. Seleção de macrófitas aquáticas com potencial de fitorremediação no arroio Santa Bárbara, município de Pelotas/RS. 2016. 52 f. TCC (Graduação)—Curso de Curso de Engenharia Ambiental e Sanitária, Universidade Federal de Pelotas, RS; 2016

[56] Dias FS et al. Aplicação de macrófitas aquáticas para tratamento de efluente doméstico. Revista Ambiental, Paraíba. 2016;**2**(1):106-115

[57] Lone MI et al. Phytoremediation of heavy metal polluted soils and water: Progresses and perspectives. Journal of Zhejiang University Science B, [s.l.]. 2008;**9**(3):210-220. DOI: 10.1631/jzus.b0710633

[58] Kawahigashi H et al. Phytoremediation of metolachlor by transgenic Rice plants expressing human CYP2B6. Journal of Agricultural and Food Chemistry, [s.l.]. 2005;**53**(23):9155-9160. DOI: 10.1021/jf0511143

[59] Ali H, Khan E, Sajad MA. Phytoremediation of heavy metals—Concepts and applications. Chemosphere, [s.l.]. 2013;**91**(7): 869-881, maio. DOI: 10.1016/j.chemosphere. 2013.01.075

[60] Delaqua GCG. Incorporação da biomassa seca de *Salvinia auriculata* aubl. proveniente do processo de fitorremediação em cerâmica vermelha. RJ, Brazil: Universidade Estadual do Norte Fluminense—vUENF campos dos Goytacazes, maio; 2016

[61] Rolon AS, Homem HF, Maltchik L. Aquatic macrophytes in natural and managed wetlands of Rio Grande do Sul State, Southern Brazil. Acta Limnologica Brasiliensia, [s.l.]. 2010;**22**(02):133-146. DOI: 10.4322/actalb.02202003

[62] de Sena Kafer D et al. Composição florística e fitossociologia de macrófitas aquáticas em um banhado continental em Rio Grande, RS, Brasil. Rodriguésia, [s.l.]. 2011;**62**(4):835-846, dez. DOI: 10.1590/s2175-78602011000400011

[63] Ting WHT, Tan IAW, Salleh SF, Wahab NA. Application of water hyacinth (Eichhornia crassipes) for phytoremediation of ammoniacal nitrogen: A review. Journal of Water Process Engineering. 2018;**22**:239-249. DOI: 10.1016/j.jwpe.2018.02.011

Chapter 11

Performance of Anoxic-Oxic Sequencing Batch Reactor for Nitrification and Aerobic Denitrification

Juan C. Alzate Marin, Alejandro H. Caravelli and Noemí E. Zaritzky

Abstract

The biological nitrogen removal (BNR) involves two processes: nitrification and denitrification. Denitrification occurs almost exclusively under facultative anaerobic or microaerophilic conditions; however, aerobic denitrification can occur in aerated reactors. In this chapter, the feasibility of achieving nitrogen removal using a lab-scale biological sequencing batch reactor (SBR) exposed to anoxic/oxic (AN/OX) phases is described in order to attain aerobic denitrification. The SBR was fed with acetate and ammonium sulfate. Nitrite generation was controlled in order to avoid the N_2O production by nitrifier denitrification. Experiments under four different operating conditions were carried out: low and high aeration, each one with low and high organic loads. For all the tested conditions, a complete COD removal was achieved. The highest inorganic N removal close to 80% was obtained at pH = 7.5, high organic load (880 mg COD/(L day)) and high aeration given by 12 h cycle, AN/OX ratio = 0.5:1.0, and dissolved oxygen concentration higher than 4.0 mg O_2/L. Nitrification followed by high-rate aerobic denitrification took place during the aerobic phase. Denitrification took place mainly from the intracellular reserves of polyhydroxyalkanoates (PHA) during the aerobic phase. The proposed AN/OX system constitutes a simple and potentially eco-friendly process for biological nitrogen removal, providing N_2 as the end product and decreasing the formation of N_2O, a powerful greenhouse gas.

Keywords: nitrogen removal, sequencing batch reactor, nitrification, aerobic denitrification, polyhydroxyalkanoates, glycogen

1. Introduction

The biological removal of nitrogen (N) comprises two processes: nitrification and denitrification. The nitrification is a strict aerobic process that involves the oxidation of ammonia (NH_3) to nitrate (NO_3^-) by autotrophic bacteria. Firstly, ammonia is oxidized to nitrite (NO_2^-), by means of ammonia-oxidizing bacteria (AOB), and then nitrite is oxidized to nitrate by the nitrite-oxidizing bacteria (NOB) [1]. In the second step, named denitrification, nitrate is converted into a gaseous product, nitrous oxide (N_2O) or molecular nitrogen (N_2), which is finally eliminated into the atmosphere. Denitrification is an anoxic process performed by heterotrophic bacteria using nitrite and/or nitrate as the electron acceptor. In full denitrification, NO_3^- is reduced to NO_2^- and then to nitric oxide (NO), N_2O, and finally to N_2 [2].

Nitrosomonas is the most common genus of autotrophic bacteria capable of oxidizing ammonium to nitrite; however, *Nitrosococcus*, *Nitrosospira*, *Nitrosovibrio*, and *Nitrosolobus* also have that ability. These ammonium oxidizers belong to the beta subdivision of the *Proteobacteria* [3]. *Nitrobacter*, *Nitrospira*, *Nitrospina*, *Nitrococcus*, and *Nitrocystis* are known to be involved in the nitrite oxidation [3]. Nitrite-oxidizing genera belong to the alpha, gamma, and delta subdivisions of the *Proteobacteria* [4]. Denitrification is carried out by several bacterial genera such as *Achromobacter*, *Aerobacter*, *Alcaligenes*, *Bacillus*, *Brevibacterium*, *Lactobacillus*, *Micrococcus*, *Proteus*, *Pseudomonas*, and *Spirillum* [5].

Carbon is not a difficult compound to eliminate by biological processes; on the contrary, one of the most common problems in wastewater treatment plants is the lack of organic carbon to carry out the denitrification process. Particularly, treatment plants with low chemical oxygen demand/nitrogen (COD/N) ratios exhibit difficulties for nitrogen removal due to a shortage of organic substrate [6, 7].

Several biological processes have been proposed for nitrogen removal. The modified Ludzack-Ettinger process is a widespread conventional technology for nitrogen biological removal. This process is a modification of a conventional activated sludge process where an anoxic reactor is located upstream of the aerobic reactor. This process with pre-anoxic configuration is commonly named anoxic/oxic (AN/OX) process. In the first reactor, denitrification is carried out using organic carbon from wastewater. For this, the process requires an internal recycle that carries nitrate, generated from ammonia by the nitrification process (aerobic reactor), to the anoxic reactor. The amount of nitrate removed in the anoxic reactor depends on both the recycle flow and availability of influent organic carbon. Several disadvantages are associated with this process: (a) high costs involved in the recirculation; (b) production of nitrogen oxides as end products, instead of N_2, which is caused by microaerophilic conditions, generated by recirculation [8]; and (c) limitation of the carbon source in the anoxic tank, caused by the recirculation of the nitrate-rich mixed liquor, resulting in accumulation of intermediate products such as nitrites and nitrogen oxides [9].

Systems based on postanoxic denitrification have the anoxic tank located downstream of the aerobic tank. Nitrification and consumption of the organic carbon take place in the first reactor. Denitrification is carried out in the anoxic stage. Thus, mixed liquor recycle from the aerobic to the anoxic stage is not required. However, this oxic/anoxic (OX/AN) system leads usually to a total consumption of the organic carbon. This configuration was firstly proposed by Wuhrmann [10], where organic substrates required for denitrification were probably supplied from endogenous death and lysis of active biomass [11]. Then, Wuhrmann process was modified to improve denitrification by carbon addition [11]. However, additional operational costs are caused by the addition of exogenous carbon such as methanol or acetate [12]. Another disadvantage is attributed to the postanoxic denitrification process. Microaerophilic conditions generated from the transfer of oxygen by mixing in the anoxic reactor can exert an inhibitory effect on the denitrification rate [13]. This phenomenon can finally trigger the production of nitrogen oxides due to incomplete denitrification.

Three main routes for biological production of N_2O have been proposed: hydroxylamine oxidation and nitrifier denitrification, both processes by AOB, and heterotrophic denitrification by heterotrophic denitrifiers [14]. N_2O emissions from heterotrophic denitrification can occur under microaerophilic conditions, because oxygen could inhibit the activity of nitrous oxide reductase [15]. At low DOC, N removal takes place via partial nitrification, and formed nitrite is denitrified to N_2/N_2O by AOB [16].

Simultaneous nitrification and denitrification (SND) are an alternative process to the conventional configurations previously described. The SND process is carried out in a single reactor where partial nitrification, from ammonia to nitrite, coupled to denitrification, takes place. SND process is based on gradients of dissolved oxygen (DO) within the flocs. The nitrifying autotrophic bacteria are distributed on the periphery of the floc, where the dissolved oxygen concentration (DOC) is above 2 mg O_2/L, while the denitrifying bacteria are located inside the floc, where the concentration of oxygen is very low [17, 18]. Large flocs (>125 μm) allow generating an oxygen gradient with anoxic conditions in the center of the floc [19, 20]. SND can be accomplished at low DOC [21]. However, at concentrations of about 0.4 mg O_2/L, N_2O instead of N_2 may be the final product of denitrification [22]. In addition, nitrite accumulation above 1 mg/L triggers the production of N_2O, and at higher nitrite levels, the denitrification process could be inhibited [21].

Another alternative process to the conventional nitrification-denitrification is based on shortcut nitrification (nitritation) followed by denitritation. In this process, AOBs oxidize NH_4^+ to NO_2^-, and then, the formed NO_2^- is denitrified [23]. Nitrogen elimination via nitrite requires high ammonia concentration and low DOC (<0.4 mg O_2/L) in order to prevent NOB growth [24]. In this process, oxygen consumption (aerobic phase) and organic carbon demand (anoxic stage) are reduced 25 and 40%, respectively, in comparison to the conventional nitrification-denitrification [25]. However, NO_2^- accumulated after nitritation is considered a key factor that triggers the N_2O generation by means of the nitrifier denitrification in a low DO environment [26]. Partial nitritation/anammox was proposed 20 years ago as key strategy for achieving a more sustainable treatment of municipal wastewater. Partial nitritation/anammox is an autotrophic nitrogen removal process based on two successive processes: partial oxidation of ammonium to nitrite by AOBs followed by oxidation of the residual ammonium with the formed nitrite to nitrogen gas [27]. The last process named anammox is carried out by a group of *Planctomycete* bacteria, which grow with CO_2 as the sole carbon source and use nitrite as the electron donor [3]. Partial nitrification, which occurs usually at low DO conditions (involving lower energy demands), can lead to NO_2^- accumulation. Nitrifier denitrification, in the presence of NO_2^- and low DO, has been considered the most likely pathway of production of N_2O in both nitritation reactor and anammox reactor [23].

Advanced N-removal processes such as partial nitrification-denitrification (shortcut nitrification, nitritation, followed by denitritation), SND, or partial nitritation-anammox are applied with the view to reducing the energy demands. However, N_2O emissions still occur and can even be higher than the ones observed during conventional nitrification-denitrification [23].

Aerobic denitrification is an alternative process to conventional anoxic denitrification, which can achieve complete denitrification at high oxygen concentrations. This process constitutes a good strategy to diminish N_2O emissions [28]. A total of 37 species (14 genera) has been reported as potential aerobic denitrifiers, which belong mainly to α, β, and γ *Proteobacteria* [29]. *Citrobacter diversus* [30], *Alcaligenes faecalis* [31], *Pseudomonas aeruginosa* [32], *Microvirgula aerodenitrificans* [33], *Paracoccus denitrificans* [32], and *Bacillus licheniformis* [34], among others, have been reported to be able to carry out aerobic denitrification. Ji et al. [29] have proposed that nitrate and oxygen co-respiration is a microbial adaption that allows the degradation of toxic nitrate in an aerobic environment. Aerobic denitrification can be an auxiliary pathway next to aerobic respiration [35]. It has been suggested that the enzymatic system for aerobic and anaerobic denitrification is probably the same. Anaerobic denitrification is negatively affected by aerobic conditions, being widely accepted that nitrous oxide reductase is inhibited by oxygen. However, N_2 generation as final product under high oxygen concentrations suggests the probable existence of different nitrous oxide reductases, which are insensitive to oxygen [35]. Denitrification via nitric oxide dismutation has been also proposed. In this process, denitrification of nitrate and nitrite to nitric oxide is followed by dismutation of nitric oxide into oxygen and N_2, which did not require nitrous oxide reductase. However, it still needs to be investigated if nitric oxide dismutation is a common and widespread process between bacteria [35].

The organic carbon required for denitrification has been considered the critical element in conventional nitrogen removal processes [36]. Therefore, it is crucial to achieve a nitrogen removal process using completely the organic carbon from wastewaters. Intracellular carbon such as PHA (polyhydroxyalkanoates) and/or glycogen is commonly stored in wastewater treatment systems. These carbon reserves could drive denitrification. Anaerobic/oxic (ANA/OX) configuration can enrich two kinds of organisms: polyphosphate-accumulating organisms (PAOs) and glycogen-accumulating organisms (GAOs) [37]. PAOs and GAOs are able to store PHA and glycogen. Denitrifying PAOs and denitrifying GAOs are able to denitrify using PHA and/or glycogen as carbon source.

The sequential batch reactor (SBR) is one of the main technologies for the biological treatment of wastewaters, being successfully used in urban wastewater [38, 39], as in industrial wastewaters [40, 41]. A SBR with anaerobic/oxic/anoxic configuration (ANA/OX/AN SBR) has been used for the removal of carbon and nitrogen. Efficient nitrogen removal via nitrification followed by post-denitrification, without the addition of external organic carbon, was reported. For this, PHA and glycogen stored during the anaerobic phase were later used as electron donors during post-anoxic denitrification. Denitrification was attributed to denitrifying glycogen-accumulating organisms [36].

In this chapter, a nitrogen removal process based on nitrification-aerobic denitrification was proposed. An anoxic/oxic (AN/OX) SBR with DOC higher than 1.5 mg O_2/L during the aerobic period was utilized. In this system, two requirements must be met: (a) growth of denitrifying bacteria able to store internally sufficient carbon reserves (PHA and/or glycogen) in the anoxic phase and (b) ability of the denitrifying bacteria to denitrify during the aerobic phase by using the intracellular carbon reserves. The AN/OX SBR would avoid both mixed liquor recirculation and exogenous carbon addition, and additionally potential emissions of N_2O could be minimized. Thus, the proposed system offers important advantages with respect to both conventional nitrification-denitrification and advanced N-removal processes.

2. Activated sludge reactor

A lab-scale SBR (1.2 L working volume) was operated for 10 months. The SBR was inoculated with sludge from a lab-scale activated sludge plant in Center of Research and Development in Food Cryotechnology (CIDCA,

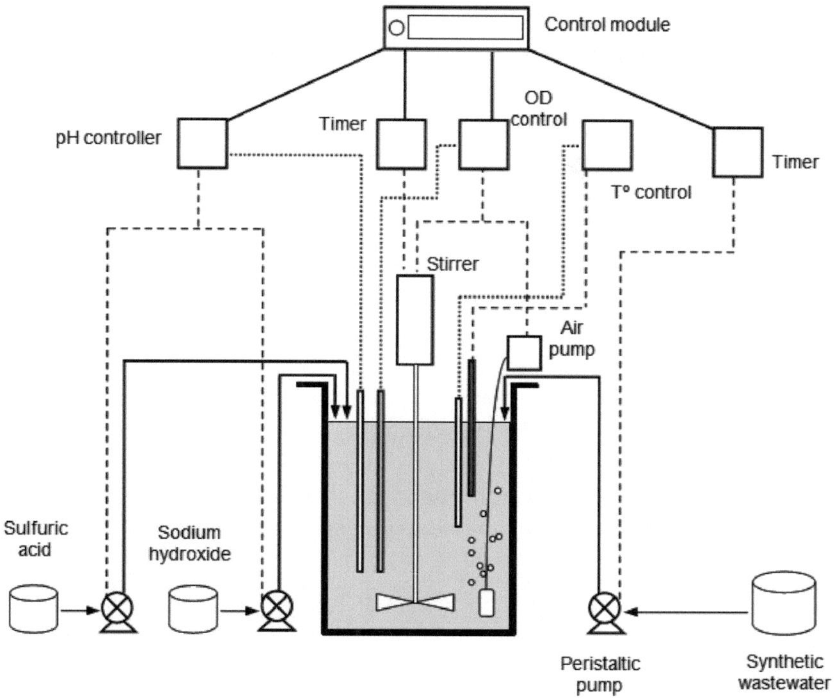

Figure 1.
Scheme of the lab-scale sequencing batch reactor (from Alzate Marin et al. [42]).

UNLP-CONICET-CIC, Argentina). The SBR was operated with cycles comprising the following phases: reaction (with anoxic and aerobic stages), biomass settling, and supernatant draw. The reactor was completely mixed at a stirring rate of 100 rpm, except during the settle and draw periods. The reactor was automatically controlled by a data acquisition and control system (DACS) developed in the electronic laboratory of CIDCA; pH was measured by a pH probe (Phoenix, Houston, TX, USA). Air was introduced through porous diffusers at the bottom of the reactor. Dissolved oxygen concentration was measured by a DO probe (Ingold Mettler Toledo, Urdorf, Switzerland) and expressed as percentage of the oxygen saturation level (OSL) by the DACS. The SBR scheme is shown in **Figure 1**.

3. Volumetric oxygen transfer coefficient

Oxygen is known to increase the oxidative state of biological systems, which could negatively affect anaerobic and anoxic processes. Microaerophilic conditions can be caused by stirring. The volumetric oxygen transfer coefficient (k_La, h^{-1}) is an important parameter in the aerobic wastewater treatment, particularly when anaerobic or anoxic conditions are required. In the present study, k_La was determined in order to evaluate the oxygen amount supplied to the reactor by agitation during the anoxic phase. k_La was measured by the clean water non-steady-state method [43] at 20°C, agitation rate of 100 rpm, and different aeration rates (vvm = 12–137 L/(L h)). Firstly, the SBR (1.2 L) was continuously aerated until the saturation concentration of oxygen (DOC*, mg O_2/L) in water was reached. Then, DO is completely removed by the addition of sodium sulfite. Finally, the aeration was turned on to the oxygen saturation level. DOC was measured at several points during the aeration period. k_La in the reactor was measured by integration of the following equation:

$$\frac{dDOC}{dt} = k_L a \ (DOC^{*} - DOC) \qquad (1)$$

where DOC* is the saturation concentration of oxygen in water (mg O_2/L) at the working temperature and DOC is the dissolved oxygen concentration (mg O_2/L) at time (t). The driving force of the oxygen transfer process is given for the difference between DOC* and DOC.

A linear relationship between k_La and the aeration rate has been proposed by the following expression:

$$k_L a = m \ AER + n \qquad (2)$$

where AER is the aeration rate (L/(L h)), m is the slope (L/L), and n (h^{-1}) corresponds to the k_La produced by stirring without aeration (AER = 0). The parameters m and n were determined through linear regression analysis (Sigma Plot 10.0) resulting in 0.10 L/L and 2.34 h^{-1}, respectively.

For clean water, at working conditions of the SBR, 25°C, stirring rate of 100 rpm, and without aeration, a k_La value of 2.63 h^{-1} was estimated by using the following expression [43]:

$$k_L a_{(25°C)} = k_L a_{(20°C)} \ 1.024^{(25-20)} \qquad (3)$$

Based on this estimation, it was assumed that only stirring will cause oxygen penetration through liquid surface during the anoxic stage of the SBR operation.

4. Synthetic wastewater and operating conditions

Synthetic wastewater (SWW) contained sodium acetate (carbon and energy source), ammonium sulfate (nitrogen source), and potassium phosphate (phosphorus source). A micronutrient solution (1 ml) was added to the SWW (1 L) [44]; influent COD/N/P ratio was 100:10:5. SWW was fed to the reactor in the first 2 min of the anoxic period. Mixed liquor was withdrawn at the end of the aerobic phase, leading to a cellular residence time (CRT) of 10 days. Treated wastewater was removed from the SBR after settling period. A volumetric exchange ratio of about 27% was set. The effects of different operating parameters, such as DOC, organic load, cycle duration, and AN/OX ratio on the ability of nitrification and denitrification were studied.

5. Analytical methods

The SBR was monitored by determination of the following physical–chemical parameters: oxidation-reduction potential (ORP, mV), orthophosphate (PO_4^{3-}-P, mg/L), ammonia nitrogen (NH_3-N, mg/L), nitrate nitrogen (NO_3^--N, mg/L), nitrite nitrogen (NO_2^--N, mg/L), soluble COD (COD_S, mg/L), and total COD (COD_T, mg/L). The oxidation-reduction potential is a measure of the oxidative state in an aqueous system. ORP reflects the concentration of DO, organic substrate, activity of organisms, and some toxic compounds in the system, the DOC being the most important factor [45]. The ORP of the SBR was measured off-line using an ORP probe (Phoenix, Houston, TX, USA). The other physical-chemical parameters were determined by spectrophotometric methods using commercial reagents (Hach Company, Loveland, CO). COD_S corresponded to the organic substrate. Biomass concentration was determined as COD (COD_B, mg/L) from the difference between COD_T and COD_S. COD_B was correlated with volatile suspended solids (VSS, mg/L).

Intracellular poly-P and PHA granules were detected by Neisser and Sudan Black staining, respectively [46]. Total carbohydrate (TC) content was determined following a modified version of the anthrone method proposed by Jenkins et al. [47].

6. Inorganic nitrogen removal

Inorganic nitrogen (Ni) corresponded to the sum of ammonia, nitrite, and nitrate concentrations. The inorganic nitrogen removal (NiR) was measured throughout the operational cycle as follows:

$$\% \text{ NiR} = \left(1 - \frac{Ni_t}{Ni_O}\right) 100 \tag{4}$$

where Ni_O is the Ni concentration at the start of the anoxic phase (mg/L) given by the NH_3-N from the wastewater and Ni_T corresponds to the Ni concentration (mg/L) at time t of the SBR operational cycle. Residual nitrate and nitrite (from of the previous cycle) were not considered in the determination of Ni_O.

7. Simultaneous nitrification-denitrification (SND) followed by denitrification (DN)

Simultaneous nitrification and denitrification (SND) took place from the beginning of the aerobic phase until the moment when the ammonium was exhausted. Later, subsequent nitrogen removal occurred by denitrification (DN).

Nitrogen removed via SND was determined in the aerobic phase from the difference between the amounts of oxidized ammonia nitrogen (NH_3-$N_{oxidized}$) and oxidized nitrogen (NO_x^--N: NO_3^--N + NO_2^--N). For SND determination, NH_3-$N_{oxidized}$ was calculated from the difference between the total NH_3-N consumption and NH_3-N assimilated into heterotrophic biomass (NH_3-$N_{assimilated}$). Nitrogen assimilated by nitrifying bacteria was assumed to be negligible [48]. The total consumption of NH_3-N was determined by spectrophotometry. NH_3-N assimilated into heterotrophic biomass was estimated for the aerobic period in the presence of ammonium. For this, theoretical mass balances of carbon and nitrogen were carried out using typical values for stoichiometric coefficients of the studied biological process. In SBR with feast-famine regime, PHB (polyhydroxybutyrate) is synthetized from acetate under anaerobic or anoxic phase, and then biomass is produced during the aerobic phase from stored PHB. In our system, PHB production was estimated using a yield $Y_{PHB/Acetate}$ of 0.52 C-mol PHB/C-mol Ac for anoxic condition [49]. Available acetate for PHB synthesis was estimated from difference between initial COD and COD required for anoxic denitrification using a stoichiometric coefficient of 3.8 mg COD_{Ac}/mg NO_3^--N. Biomass production from PHB was estimated assuming a heterotrophic biomass yield $Y_{X/PHB}$ of 0.5 C-mol X/C-mol PHB. Finally, NH_3-$N_{assimilated}$ by heterotrophs was determined assuming a biomass molecular formula of $CH_{1.8}O_{0.5}N_{0.2}$, which is equivalent to 24.6 g VSS/C-mol X [48].

SND was calculated from the following equation [48]:

$$\% \text{ SND} = \left(1 - \frac{NO_X^- - N}{NH_3 - N_{oxidized}}\right) 100 \tag{5}$$

where NO_x^--N is the sum of the oxidized nitrogen species (nitrite and nitrate) at the moment when ammonia was exhausted and NH_3-$N_{oxidized}$ corresponds to the ammonia nitrogen oxidized during the aerobic period.

Nitrogen removed via denitrification (DN) was calculated from the difference between oxidized nitrogen at the end of nitrification (NO_x^--N_{FN}) and oxidized nitrogen at the end of the aerobic phase (NO_x^--N_{FA}) as follows:

$$\% \, DN = \left(1 - \frac{NO_x^- - N_{FA}}{NO_x^- - N_{FN}}\right) 100 \quad (6)$$

8. Experimental design

Experiments were carried out at low and high dissolved oxygen concentrations (oxygen saturation levels, OSL, of 20 and 60%, respectively) using in each case low and high organic loads (440 and 880 mg COD/L day). The following notation was used to describe and report the results of the experiments: low oxygen concentration and low organic load (LOLC), low oxygen concentration and high organic load (LOHC), high oxygen concentration and low organic load (HOLC), and high oxygen concentration and high organic load (HOHC).

9. Experiments at low dissolved oxygen concentrations

In these experiments, the effect of organic load on the nitrification process was evaluated at low DOC. An oxygen saturation level (OSL) of 20%, equivalent to a DOC of 1.6 mg O_2/L, was set for the aerobic phase (**Table 1**). Experiments were carried out at two different organic volumetric loads. In experiment low oxygen concentration and low organic load (LOLC), 440 mg COD/(L day) was used, and in experiment low oxygen concentration and high organic load (LOHC), the value was 880 mg COD/(L day).

In the experiments LOLC, the SBR showed at steady state a good performance with a biomass concentration of 1220 ± 215 mg COD_B/L. For organic carbon, a removal higher than 99% was reached in anoxic phase. Ammoniacal nitrogen removal was about 99%, mainly in the aerobic phase (**Figure 2**). In this phase, about 70% of the ammonium was nitrified up to nitrate as was determined by mass balance. According to these results, a redox potential of about +295 mV was measured during the aerobic phase, which involves a suitable oxidizing environment for autotrophic nitrification. It must be considered that ORP values between +100 and +350 mV are necessary for the nitrification process to take place [50]. A relatively low concentration of oxygen (<2.0 mg O_2/L) was enough to achieve a good nitrifying activity without accumulation of nitrite. Volumetric and specific nitrification rates are shown in **Table 2**.

PHA accumulation followed by degradation of the polymer took place in the anoxic and aerobic phases, respectively, as was detected by Sudan Black staining. Cocci-shaped cells arranged in tetrads (tetrad-forming organisms, TFOs) displayed that metabolic ability (**Figure 3a** and **b**). Some subgroups of *Alphaproteobacteria* and *Gammaproteobacteria* exhibit TFO morphology with GAO phenotype. These microorganisms are commonly associated with enhanced biological phosphorus removal (EBPR) deterioration [51]. In the present study, TFOs corresponded likely to some group of GAO commonly found in systems without EBPR.

PHA could be used as intracellular carbon source for denitrification. However, poor denitrification took place since at the end of the operational cycle, the final effluent exhibited a nitrate concentration of 4.75 ± 0.25 mg NO_3^--N/L, equivalent to about 70–80% of the nitrified ammoniacal nitrogen. According to these results,

Parameters	Experiment LOLC	Experiment LOHC
Anoxic phase (min)	150	150
Aerobic phase (min)	150	150
Settling phase (min)	50	50
Draw phase (min)	10	10
Total cycle length (h)	6	6
Anoxic/aerobic ratio	1.0:1.0	1.0:1.0
Temperature (°C)	25 ± 0.5	25 ± 0.5
pH (anoxic and aerobic phases)	7.0 ± 0.1	7.0 ± 0.1
Oxygen saturation level (%)	20	20
Organic volumetric load (mg COD/(L day))	440	880
Nitrogen volumetric load (mg NH_3-N/(L day))	44	88
Phosphorus volumetric load (mg P/(L day))	22	44

Adapted from Alzate Marin et al. [42].

Table 1.
Operating conditions for experiments at low oxygen concentration.

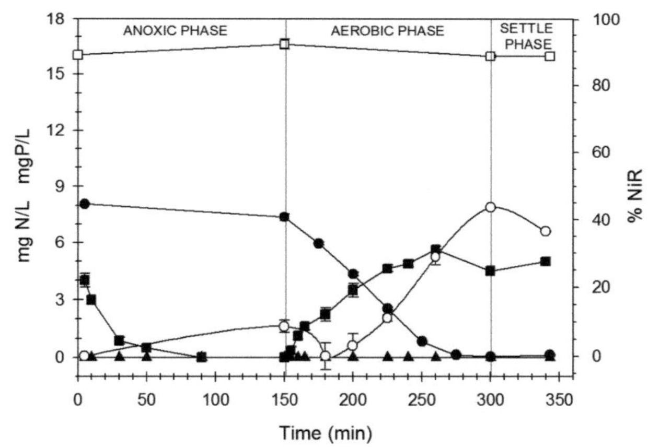

Figure 2.
Changes of phosphorus and nitrogen concentrations during operational cycles of the steady-state SBR. Experiment with low oxygen concentration and low organic load (LOLC). (□) Orthophosphate (PO_4^{3-}-P, mg P/L), (●) ammonia (NH_3-N, mg N/L), (■) nitrate (NO_3^--N, mg N/L), (▲) nitrite (NO_2^--N, mg N/L), and (○) % inorganic nitrogen removal (% NiR).

nitrogen removal through the SND and DN processes represented 11 ± 10% and 5 ± 5%, respectively. The final effluent exhibited an inorganic nitrogen concentration of 4.84 ± 0.40 mg N/L, which resulted in a mean discharge of 5.80 mg N/day. These results involved an inorganic nitrogen removal of 45 ± 2% (**Table 2**). This poor nitrogen removal was associated with the low denitrification ability of the system. It must be considered that the residual nitrate, after the discharge of the final effluent, was completely removed by denitrification in the first 90 min of the following cycle (**Figure 2**).

Poly-P staining by Neisser method resulted negative (**Figure 3c**), and soluble phosphorus (orthophosphate) concentration did not show important changes (**Figure 2**). These results involve that PAO activity and hence the EBPR process did not take place. According to these findings, positive ORP values (+286 ± 8 mV) were

Parameters	Experiment LOLC	Experiment HOLC	Experiment HOHC
VNR (mg NH_3-N/(L h))	3.96 ± 0.10	3.71 ± 0.45	4.09 ± 0.08
SNR (mg NH_3-N/(g VSS h))	4.22 ± 0.10	4.14 ± 0.48	1.33 ± 0.00
VDNR (mg NO_3^--N/(L h))	ND	2.53 ± 0.96	2.57 ± 0.36
SDNR (mg NO_3^--N/(g VSS h))	ND	2.94 ± 1.10	0.83 ± 0.10
% NAS	—	10.0 ± 1.0	28.7 ± 0.5
% SND	11 ± 10	0 ± 0	9 ± 2
% DN	5 ± 5	55 ± 3	57 ± 2
% AR	99 ± 1	99 ± 1	99 ± 1
% NiR	45 ± 2	67 ± 2	78 ± 1

Adapted from Alzate Marin et al. [42].
ND, not determined.

Table 2.
Biological parameters of the SBR for the different experiments.

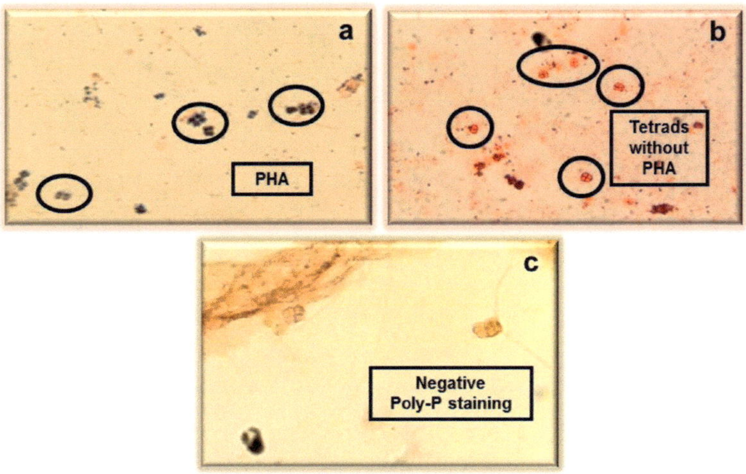

Figure 3.
Micrographs of activated sludge stained with Sudan black (a and b) and Neisser (c). (a) Tetrad-forming organisms (TFOs) showing positive PHA staining (final anoxic phase), (b) TFOs with negative PHA staining (final aerobic phase), and (c) negative Neisser staining.

recorded throughout the anoxic phase, which are not suitable for anaerobic PHA metabolism. It is well known that negative ORP values between −50 and − 200 mV are usually required for anaerobic polyphosphate breakdown [52]. In the anoxic phase, zero DOC was registered, and a k_La value of 2.63 h^{-1} was estimated by using Eq. (3). For these conditions, an oxygen transfer rate of 21.3 mg O_2/(L h) was estimated at 25°C by using Eq. (1). The oxygen transfer by stirring increased the oxidative state (positive ORP) during the anoxic phase. It can be assumed that this phenomenon would lead to unfavorable ecological conditions for anaerobic metabolism of PAOs.

In the experiments with low oxygen concentration and high organic load (LOHC), the organic volumetric load was increased from 440 to 880 mg COD/(L day) under identical operational conditions to those of the experiment LOLC (**Figure 4**). The nitrogen and phosphorus volumetric load were 88 mg

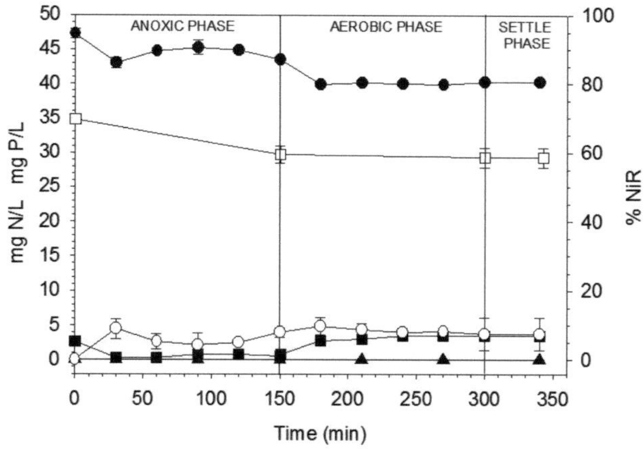

Figure 4.
Changes of phosphorus and nitrogen concentrations during operational cycles of the steady-state SBR. Experiment with low oxygen concentration and high organic load (LOHC). (□) Orthophosphate (PO_4^{3-}-P, mg P/L), (●) ammonia (NH_3-N, mg N/L), (■) nitrate (NO_3^--N, mg N/L), (▲) nitrite (NO_2^--N, mg N/L), and (○) % inorganic nitrogen removal (% NiR) (adapted from Alzate Marin et al. [42]).

NH3-N/(L day) and 44 mg P/(L day), respectively, in order to maintain the same COD/N/P ratio (100:10:5) (**Table 1**). The steady-state SBR reached a biomass concentration of 1850 ± 120 mg COD_B/L. Ammoniacal nitrogen was removed only 15% throughout the operational cycle. Poor nitrification was observed as only 7% of ammonia from anoxic phase was nitrified, even though adequate oxidizing conditions were registered during the aerobic phase (ORP > +100 mV). Low nitrate concentrations were generated, and hence the denitrification process did not take place; nitrite was not accumulated. The final effluent showed a high inorganic nitrogen concentration (43.5 ± 0.20 mg N/L), resulting in a mean discharge of 57.42 mg N/day. Thus, a poor Ni removal of only 8% was achieved (**Figure 4**). It is important to highlight that even though the influent nitrogen load was only two times higher to that of the experiment LOLC, the daily nitrogen discharge was about ten times greater than that corresponding to the previous assay. EBPR activity was not observed; as was previously discussed for experiment LOLC, oxidizing conditions during the anoxic phase (positive ORP) were unfavorable for PAO growth.

In the tested system, a COD/N/P ratio of 100:10:5 was utilized in experiments LOLC and LOHC in order to ensure excess conditions of nitrogen and phosphorus. Nevertheless, a relatively low DO concentration was used, which can lead to competition between heterotrophic and nitrifying bacteria. In the experiment LOHC, the higher organic load led to a greater intracellular PHA production, in anoxic phase, in comparison to LOLC. Thus, a higher growth of heterotrophic bacteria from PHA took place in the aerobic phase, which would involve a greater oxygen uptake rate by heterotrophs. This observation was reported by Third et al. [48] working with an aerobic SBR fed with acetate. Nitrifying bacteria, with very low growth rate, were likely outcompeted by heterotroph overgrowth under low oxygen availability. This phenomenon could explain the poor nitrifying activity in experiment LOHC. In conclusion, the organic load stimulated strongly the competition by oxygen between heterotrophic and nitrifying bacteria at low DO concentrations.

10. Experiments at high dissolved oxygen concentration

In these assays, at high dissolved oxygen concentration, a value of OSL (60%), equivalent to a DOC of 4.8 mg O_2/L, was set for the aerobic phase (**Table 3**). As in

Parameters	Experiment HOLC	Experiment HOHC
Anoxic phase (min)	220	220
Aerobic phase (min)	440	440
Settling phase (min)	51	51
Draw phase (min)	9	9
Total cycle length (h)	12	12
Anoxic/aerobic ratio	0.5:1.0	0.5:1.0
Temperature (°C)	25 ± 0.5	25 ± 0.5
pH (anoxic and aerobic phases)	7.5 ± 0.1	7.5 ± 0.1
Oxygen saturation level (%)	60	60
Organic volumetric load (mg COD/(L day))	440	880
Nitrogen volumetric load (mg NH_3-N/(L day))	44	44
Phosphorous volumetric load (mg P/(L day))	22	44

Table 3.
Operational conditions for experiments at high dissolved oxygen concentration with different organic loads.

the previous experiments, two organic volumetric loads were evaluated: 440 and 880 (mg COD/(L day)) (**Table 3**). The effects of cycle duration, anoxic/aerobic ratio, and organic load on the denitrification process were evaluated. The purpose of these experiments was to determine optimal experimental conditions to attain a good denitrifying activity and hence an acceptable process of nitrogen removal. Therefore, in addition to achieving efficient nitrification, sufficient organic carbon must be supplied for the denitrification process to take place. High oxygen availability permitted to minimize competition by oxygen between heterotrophic and nitrifying bacteria. In these experiments, the extension of the operating cycle was increased from 6 h to 12 h, and the anoxic/aerobic ratio was decreased from 1.0:1.0 to 0.5:1.0. These conditions were set in order to provide a longer aerobic period to favor the denitrification process.

In the experiment HOLC, the volumetric loads of organic carbon, nitrogen, and phosphorus were the same as those used in the experiment LOLC. All the operating conditions are shown in **Table 3**.

The COD/N/P ratio (100:10:5) and oxygen saturation level (60%) used in this assay would minimize competition between heterotrophs and nitrifiers. Oxidizing conditions were registered in the anoxic phase (ORP = +187 ± 13), being unfavorable for the EBPR process to occur. Ammonium was almost completely removed (99%). About 80% was eliminated in the aerobic phase. Nitrification produced nitrate concentrations of about 10–12 mg NO_3^--N/L in the first 2 h of the aerobic period. ORP values higher than +190 mV favored the nitrifying activity. Then, the nitrate concentration gradually decreased, which was attributed to the activity of denitrifying bacteria (**Figure 5**). The mean discharge of nitrate was 3.2 mg N/day. This concentration was about 32% lower than the one obtained in experiment LOLC for a same nitrogen volumetric load.

Residual nitrate was denitrified at the beginning of the following cycle (anoxic phase). Nitrite was not accumulated in the SBR, as was also observed in the previous experiments. The mean discharge of inorganic nitrogen was 3.2 mg N/day (corresponding totally to nitrate), being about 45% lower than the results obtained in experiment LOLC. According to the nitrogen mass balance, about 85% of the incoming ammonia in aerobic period was nitrified; nitrogen assimilation by heterotrophic bacteria corresponded to 15%. Nitrogen assimilated by heterotrophs represented

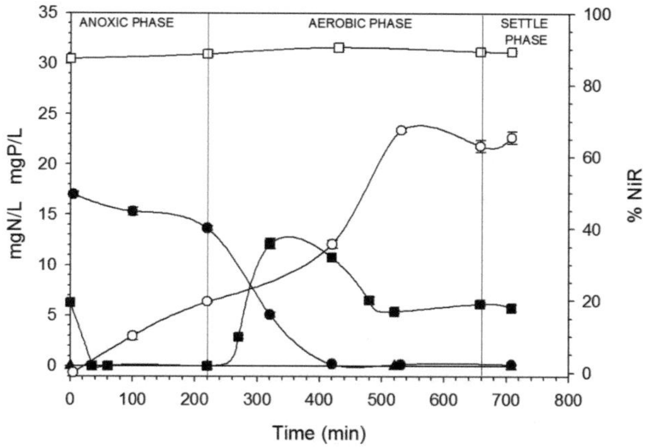

Figure 5.
Changes of the phosphorus and nitrogen concentrations during an operational cycle of the steady-state SBR (experiment HOLC). (□) orthophosphate (PO_4^{3-}-P, mg P/L), (●) ammonia (NH_3-N, mg N/L), (■) nitrate (NO_3^--N, mg N/L), (▲) nitrite (NO_2^--N, mg N/L), and (○) % inorganic nitrogen removal (% NiR) (adapted from Alzate Marin et al. [42]).

10% of the total ammonia load applied to the SBR. Volumetric and specific nitrification rates were not significantly different to those determined in the experiment LOLC. SND did not take place; denitrification began once the nitrification process was completed; 55 ± 3% of the generated nitrate was removed (**Table 2**).

Nitrification followed by denitrification was the most important process for nitrogen removal. The elimination of Ni was about 50% higher than that achieved in experiment LOLC (**Table 2**). The greater efficiency for nitrogen removal was attributed to a higher denitrifying activity in the experiment HOLC. In addition, the improved denitrification process of this assay can be attributed to a greater extension of the aerobic phase. However, the denitrification was probably limited by a low availability of intracellular organic carbon during the aerobic phase. In the experiment HOHC, the organic volumetric load was increased from 440 to 880 mg COD/(L day), while the ammoniacal nitrogen load was the same as that corresponding to the HOLC (44 mg NH_3-N/(L day)). This led to an increase in the COD/N ratio from 100:10 to 100:5. The volumetric load of phosphorus was 29 mg P/(L day). The other operating conditions were identical to those used in the experiment HOLC (**Table 3**).

Organic substrate was completely removed in anoxic phase. Ammonium was almost depleted during the process; about 80–85% was eliminated in the aerobic phase (**Figure 6**). Nitrogen assimilated by heterotrophs represented almost 30% of the incoming ammonia to the SBR (**Table 2**). Oxidizing conditions were similar to those corresponding to previous assays, with positive ORP values. The specific nitrification rate was significantly lower than that corresponding to the assay HOLC (**Table 2**). This result was attributed to the enrichment of the biomass in heterotrophic bacteria because of the higher organic load applied in experiment HOHC. Biomass concentration was twice the value reached in the HOLC assay.

The SND process showed little improvement. The denitrification was similar to that obtained in experiment HOLC, and the specific denitrification rate was significantly lower than that observed in the previous experiment. The mean discharge of inorganic nitrogen was 2.2 mg N/day. The inorganic nitrogen removal was 78 ± 1%, being significantly higher than that observed in the previous assay (**Table 2**). In the experiments HOHC, the higher organic load generated a greater PHA production, as was estimated by material balance, in comparison with HOLC assay. Thus, a higher content of endogenous carbon and energy reserve for the denitrification

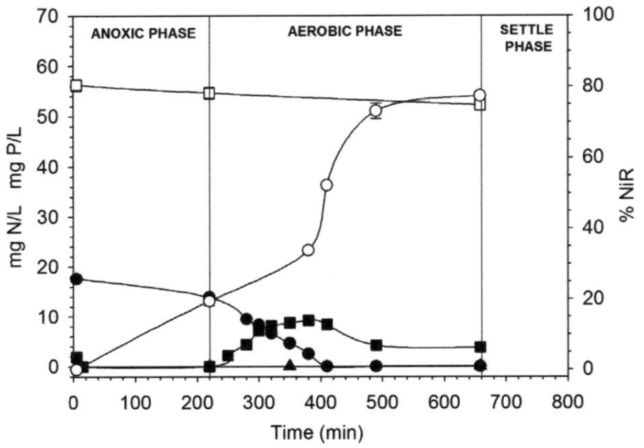

Figure 6.
Changes of the phosphorus and nitrogen concentrations during an operational cycle of the steady-state SBR (experiment HOHC). (□) orthophosphate (PO_4^{3-}-P, mg P/L), (•) ammonia (NH_3-N, mg N/L), (■) nitrate (NO_3^--N, mg N/L), (▲) nitrite (NO_2^--N, mg N/L), and (○) % inorganic nitrogen removal (% NiR).

process was available. However, the higher efficiency of inorganic nitrogen removal attained in experiment HOHC was attributed mainly to a greater assimilation of nitrogen by heterotrophic bacteria, which was about three times larger than that observed at low organic load (**Table 2**).

As was mentioned, the highest inorganic nitrogen removal was attained in the experiments HOHC; however, the specific denitrification rate was significantly lower than that corresponding to the assay HOLC. It must be considered that a high organic load led to an excessive growth of heterotrophs, which probably involved an intense competition by different growth factors among heterotrophic bacteria. Under these conditions, it can be inferred that denitrifying bacteria would preferably use oxygen as the final acceptor of electrons instead of nitrate, which represents a competitive advantage in terms of energy efficiency. This would explain the low specific denitrification rate obtained in the HOHC experiment.

11. Endogenous carbon sources as affecting microbial consortia in denitrification process

In all the experiments, the denitrification process at aerobic phase took place without external organic carbon. Denitrification occurred from intracellular carbon and energy reserves; the specific denitrification rates obtained were higher than those corresponding to endogenous decay (0.2–0.6 mg NO_3^--N/(g VSS h) [53]). Under steady-state conditions, the total carbohydrate (TC) concentration of the biomass was determined by the anthrone method throughout the operational cycle of the reactor. TC increased slightly during the anoxic phase and initial period of the aerobic phase, and then it decreased slightly at the end of the aerobic phase. These TC changes could not be attributed to cyclic changes of intracellular glycogen, which are typical of reactors with anaerobic/aerobic regime. In these systems, the microbial community is commonly enriched with GAOs and/or PAOs, which are responsible for the degradation and synthesis of glycogen during the anaerobic and aerobic stages, respectively. In the case of GAOs, glycogen constitutes the primary source of energy for both uptake of exogenous organic carbon and PHA storage during the initial anaerobic stage [51, 54]. Then, glycogen is replenished aerobically from PHA. In the anoxic/oxic SBR of the present study, GAOs as tetrad-arranged cocci and positive PHA staining were microscopically detected. However, typical

GAO metabolism regarding glycogen cycling was not observed. TC increase was mainly attributed to microbial growth instead of glycogen accumulation, even though a light glycogen increase during the anoxic phase of the operational cycle cannot be discarded. Slight decay of TC at final aerobic phase could be attributed to the glycogen component. Anyway, GAO was not a representative microbial phenotype in the anoxic-oxic SBR. This result could be explained considering that oxidative conditions were prevalent in the anoxic period generated by the high oxygen transfer during the agitation.

Based on this analysis, it can be argued that the denitrification achieved in the SBR took place from the intracellular reserves of PHA during the aerobic phase. Denitrification process could also be driven from intracellular glycogen but to a lesser extent. PAOs and GAOs are able to denitrify using intracellular carbon source. In the present study, PAO activity was not observed. The absence of EBPR activity was associated to high oxidative conditions not favorable to PAOs during anoxic phase more than to the GAO-PAO competition. GAOs with tetrad-type morphology were probably responsible of the denitrification process; however, the denitrifying activity of other microbial groups should not be discarded.

The specific denitrification rates obtained in the present study were similar (experiment HOHC) or higher (experiment HOLC) than those reported in literature for anoxic denitrification carried out by PAOs; intracellular glycogen was the carbon source used for anoxic denitrification [9, 55]. Vocks et al. [56] reported a similar SDNR to that obtained in the experiment HOLC, using a membrane bioreactor (ANA/OX/AN); denitrifying GAOs were considered as responsible for the denitrification using stored glycogen as internal carbon source [56]. Li et al. [36] reported SDNRs of 0.5 and 1.24 mg NO_3^--N/(g VSS h) using glycogen and PHA, respectively, at anoxic conditions. These SDNRs were similar to that obtained in the experiment HOHC and 2–6 times lower than that corresponding to experiment HOLC.

Anoxic denitrification rates are commonly higher than those obtained under aerobic conditions [57]. In contrast, the specific denitrification rates (SDNR) obtained in the present study, at bulk DO concentration higher than 4.0 mg O_2/L, were similar or higher to those reported for anoxic conditions.

12. Conclusions

A lab-scale sequencing batch reactor (SBR) operated with two phases, anoxic and aerobic, achieved complete COD removal. At low DO concentration, the nitrification process depended on the organic load. Low DO concentration and relatively high organic load (LOHC) led to significant growth of heterotrophic bacteria and poor nitrification. At low DO concentration and low organic load (LOLC), a good nitrifying activity led to an inorganic nitrogen removal of about 45%. It is known that in activated sludge systems, competition by growth factors (macro- and micronutrients and DO) between heterotrophic and nitrifying bacteria can occur. In both experiments, LOLC and LOHC, a COD/N/P ratio of 100:10:5 assured excess conditions of nitrogen and phosphorus. Nevertheless, competition by oxygen between both groups of microorganisms took place at high organic load.

With reference to the experiments carried out at high oxygen concentration (HOLC and HOHC), a high DOC minimized competition by oxygen between heterotrophs and nitrifiers. Higher inorganic nitrogen removal (67–78%) was achieved at the following conditions: pH = 7.5, higher dissolved oxygen concentration, and prolonged aerobic phase. Nitrification followed by denitrification during the aerobic phase was the most important process for nitrogen removal. The elimination of Ni was 50–70% higher than that achieved in experiment LOLC. The greater efficiency for nitrogen removal was attributed to a higher denitrifying activity, due to a greater extension of the aerobic phase. From the results obtained using high dissolved oxygen concentrations (HOLC and HOHC), it can be concluded that there was no shortage of intracellular carbon and energy reserve. Thus, organic carbon was not the limiting substrate for the denitrification process under aerobic conditions. Denitrification took place mainly from the intracellular reserves of PHA during the aerobic phase. Aerobic denitrification could be attributed to glycogen-accumulating organism (GAOs) with tetrad-type morphology; activity of polyphosphate-accumulating organisms (PAOs) was not observed. Other microbial groups have probably contributed to the denitrifying activity. The nitrification followed by denitrification, under aerobic conditions, analyzed in the present chapter, is an alternative process to the conventional configurations. The specific denitrification rates, at bulk DO concentration higher than 4.0 mg O_2/L, were similar or higher to those reported for anoxic conditions. It is widely accepted that in an aerobic environment, denitrifying bacteria can survive in the anaerobic/anoxic center of the microbial flocs. If not, denitrifiers could tolerate oxygen so that the denitrification process is not affected. Aerobic denitrifiers can use alternatively nitrate or oxygen as final electron acceptor. In the present study, denitrifying activity was attributed to the aerobic denitrification process.

The proposed AN/OX system constitutes a simple and potentially eco-friendly process for biological nitrogen removal, providing N_2 as the end product and decreasing the formation of N_2O, a greenhouse gas that has an important influence on atmosphere warming.

Acknowledgements

The authors gratefully acknowledge the financial support given by the Consejo Nacional de Investigaciones Científicas y Técnicas (CONICET), Universidad Nacional de La Plata, and Agencia Nacional de Promoción Científica y Tecnológica, Argentina.

Nomenclature

AER	aeration rate (L/(L h)
AN	anoxic
AR	ammonia removal
ANA	anaerobic
AOB	ammonia-oxidizing bacteria
COD	chemical oxygen demand
COD_B	biomass concentration as COD (mg COD_B/L)
COD_S	soluble COD (mg/L)
COD_T	total COD (mg/L)
CRT	cellular residence time (days)
DACS	data acquisition and control system
DN	denitrification
DOC	dissolved oxygen concentration (mg O_2/L)
DOC*	saturation concentration of oxygen (mg O_2/L)
GAOs	glycogen-accumulating organisms
k_La	volumetric oxygen transfer coefficient (h^{-1})
LOHC	low oxygen concentration and high organic load

LOLC	low oxygen concentration and low organic load
HOHC	high oxygen concentration and high organic load
HOLC	high oxygen concentration and low organic load
N_2	molecular nitrogen
N_2O	nitrous oxide
NAS	nitrogen assimilated by heterotrophic bacteria
NH_3	ammonia
NH_3-N	ammonia nitrogen (mg/L)
$NH_3\text{-}N_{assimilated}$	ammonia nitrogen assimilated by heterotrophs (mg/L)
$NH_3\text{-}N_{oxidized}$	oxidized ammonia nitrogen (mg/L)
Ni	inorganic nitrogen (mg/L)
Ni_O	Ni concentration at the start of the anoxic phase (mg/L)
Ni_T	Ni concentration at time t (mg/L)
NiR	inorganic nitrogen removal
NO	nitric oxide
NO_2^-	nitrite
NO_2^--N	nitrite nitrogen (mg/L)
NO_3^-	nitrate
NO_3^--N	nitrate nitrogen (mg/L)
NO_x^--N	oxidized nitrogen (mg/L)
$NO_x^-\text{-}N_{FA}$	oxidized nitrogen at the end of the aerobic phase (mg/L)
$NO_x^-\text{-}N_{FN}$	oxidized nitrogen at the end of nitrification (mg/L)
NOB	nitrite-oxidizing bacteria
OSL	oxygen saturation level
OX	oxic
PHA	polyhydroxyalkanoates
PAOs	polyphosphate-accumulating organisms
PO_4^{3-}-P	orthophosphate (mg/L)
SBR	sequencing batch reactor
SDNR	specific denitrification rate (mg NO_3^--N/(g VSS h)
SND	simultaneous nitrification and denitrification
SNR	specific nitrification rate (mg NH_3-N/(g VSS h)
SWW	synthetic wastewater
TFOs	tetrad-forming organisms
TC	total carbohydrates
VDNR	volumetric denitrification rate (mg NO_3^--N/(L h))
VNR	volumetric nitrification rate (mg NH_3-N/(L h))
VSS	volatile suspended solids (mg VSS/L)
$Y_{PHB/Acetate}$	yield coefficient for PHB from acetate (C-mol PHB/C-mol Ac)
$Y_{X/PHB}$	yield coefficient for heterotrophic biomass from PHB (C-mol X/C-mol PHB)

Author details

Juan C. Alzate Marin[1], Alejandro H. Caravelli[1*] and Noemí E. Zaritzky[1,2]

1 Center of Research and Development in Food Cryotechnology (CIDCA), (CONICET, UNLP, CIC), La Plata, Argentina

2 Faculty of Engineering, Universidad Nacional de La Plata (UNLP), La Plata, Argentina

*Address all correspondence to: alejandrocaravelli@hotmail.com

References

[1] Zhou Y, Oehmen A, Lim M, Vadivelu V, Ng WJ. The role of nitrite and free nitrous acid (FNA) in wastewater treatment plants. Water Research. 2011;**45**:4672-4682. DOI: 10.1016/j.watres.2011.06.025

[2] Rodríguez DC, Ramírez O, Peñuela Mesa G. Behavior of nitrifying and denitrifying bacteria in a sequencing batch reactor for the removal of ammoniacal nitrogen and organic matter. Desalination. 2011;**273**:447-452. DOI: 10.1016/j.desal.2011.01.068

[3] Ahn Y-H. Sustainable nitrogen elimination biotechnologies: A review. Process Biochemistry. 2006;**41**:1709-1721. DOI: 10.1016/j.procbio.2006.03.033

[4] Teske A, Alm E, Regan JM, Toze S, Rittmann BE, Stahl DA. Evolutionary relationships among ammonia- and nitrite-oxidizing bacteria. Journal of Bacteriology. 1994;**176**:6623-6630. DOI: 10.1128/jb.176.21.6623-6630.1994

[5] Metcalf & Eddy. Ingeniería de aguas residuales. Tratamiento, vertido y reutilización. Tercera edición. New York: McGraw-Hill; 1995

[6] Pitman AR. Design considerations for nutrient removal plants. Water Science and Technology. 1991;**23**:781-790. DOI: 10.2166/wst.1991.0529

[7] Zeghal S, Puznava N, Subra JP, Sauvegrain P, Vignoles C. Methanol dosing feedback control for denitrification. In: Conference Biological Nutrient Removal. Part 2. BNR 3; December 1997; Convention Centre, Brisbane, AWA Australia. pp. 321-328

[8] Robertson LA, Dalsgaar T, Revsbech NP, Kuenen JG. Confirmation of 'aerobic denitrification' in batch cultures, using gas chromatography and ^{15}N mass spectrometry. FEMS Microbiology Ecology. 1995;**18**:113-120. DOI: 10.1111/j.1574-6941.1995.tb00168.x

[9] Winkler M, Coats ER, Brinkman CK. Advancing post-anoxic denitrification for biological nutrient removal. Water Research. 2011;**45**:6119-6130. DOI: 10.1016/j.watres.2011.09.006

[10] U.S. EPA. Manual Nitrogen Control, EPA/625/R93/010. Washington, DC: Office of Research and Development, Office of Water; 1993. 312 p

[11] Seviour RJ, Lindrea KC, Oehmen A. The activated sludge process. In: Seviour R, Nielsen PH, editors. Microbial Ecology of Activated Sludge. London, UK: IWA Publishing; 2010. pp. 57-94

[12] Tchobanoglous G, Burton FL, Stensel HD. Wastewater Engineering: Treatment and Reuse. 4th ed. McGraw Hill; 2003

[13] Plósz BG, Jobbágy A, CPL GJ. Factors influencing deterioration of denitrification by oxygen entering an anoxic reactor through the surface. Water Research. 2003;**37**:853-863. DOI: 10.1016/S0043-1354(02)00445-1

[14] Guo G, Wang Y, Hao T, Wu D, Chen G-H. Enzymatic nitrous oxide emissions from wastewater treatment. Frontiers of Environmental Science and Engineering. 2018;**12**:1-12. DOI: 10.1007/s11783-018-1021-3

[15] von Schulthess R, Wild D, Gujer W. Nitric and nitrous oxides from denitrifying activated-sludge at low-oxygen concentration. Water Science and Technology. 1994;**30**:123-132. DOI: 10.2166/wst.1994.0259

[16] Kampschreur MJ, Temmink H, Kleerebezem R, Jetten MSM, van Loosdrecht MCM. Nitrous oxide

emission during wastewater treatment. Water Research. 2009;**43**:4093-4103. DOI: 10.1016/j.watres.2009.03.001

[17] Hibiya K, Terada A, Tsuneda S, Hirata A. Simultaneous nitrification and denitrification by controlling vertical and horizontal microenvironment in a membrane-aerated biofilm reactor. Journal of Bacteriology. 2003;**100**:23-32. DOI: 10.1016/S0168-1656(02)00227-4

[18] Zhu G, Peng Y, Li B, Guo J, Yang Q, Wang S. Biological nitrogen removal from wastewater. Reviews of Environmental Contamination and Toxicology. 2008;**192**:159-195. DOI: 10.1007/978-0-387-71724-1-5

[19] Li DH, Ganczarczyk JJ. Structure of activated sludge flocs. Biotechnology and Bioengineering. 1990;**35**:57-65. DOI: 10.1002/bit.260350109

[20] Pochana K, Keller J. Study of factors affecting simultaneous nitrification and denitrification (SND). Water Science and Technology. 1999;**39**:61-68. DOI: 10.1016/S0273-1223(99)00123-7

[21] Zeng RJ, Lemaire R, Yuan Z, Keller J. Simultaneous nitrification, denitrification, and phosphorus removal in a lab-scale sequencing batch reactor. Biotechnology and Bioengineering. 2003;**84**:170-178. DOI: 10.1002/bit.10744

[22] Tallec G, Garnier J, Billen G, Gousailles M. Nitrous oxide emissions from denitrifying activated sludge of urban wastewater treatment plants, under anoxia and low oxygenation. Bioresource Technology. 2008;**99**:2200-2209. DOI: 10.1016/j.biortech.2007.05.025

[23] Massara TM, Malamis S, Guisasola A, Baeza JA, Noutsopoulos C, Katsou E. A review on nitrous oxide (N_2O) emissions during biological nutrient removal from municipal wastewater and sludge reject water. Science of the Total Environment. 2017;**596-597**:106-123. DOI: 10.1016/j.scitotenv.2017.03.191

[24] Schmidt I, Sliekers O, Schmid M, Bock E, Fuerst J, Kuenen JG, et al. New concepts of microbial treatment processes for the nitrogen removal in wastewater. FEMS Microbiology Reviews. 2003;**27**:481-492. DOI: 10.1016/S0168-6445(03)00039-1

[25] U.S. EPA. Nutrient control design manual. EPA/600/R-10/100. Cincinnati, Ohio: United States Environmental Protection Agency, Office of Research and Development; 2010. 104 p

[26] Desloover J, Vlaeminck SE, Clauwaert P, Verstraete W, Boon N. Strategies to mitigate N_2O emissions from biological nitrogen removal systems. Current Opinion in Biotechnology. 2012;**23**:474-482. DOI: 10.1016/j.copbio.2011.12.030

[27] Lackner S, Gilbert EM, Vlaeminck SE, Joss A, Horn H, van Loosdrecht MCM. Full-scale partial nitritation/anammox experiences—An application survey. Water Research. 2014;**55**:292-303. DOI: 10.1016/j.watres.2014.02.032

[28] Takaya N, Catalan-Sakairi MAB, Sakaguchi Y, Kato I, Zhou Z, Shoun H. Aerobic denitrifying bacteria that produce low levels of nitrous oxide. Applied and Environmental Microbiology. 2003;**69**:3152-3157. DOI: 10.1128/AEM.69.6.3152-3157.2003

[29] Ji B, Yang K, Zhu L, Jiang Y, Wang H, Zhou J, et al. Aerobic denitrification: A review of important advances of the last 30 years. Biotechnology and Bioprocess Engineering. 2015;**20**:643-651. DOI: 10.1007/s12257-015-0009-0

[30] Huang HK, Tseng SK. Nitrate reduction by *Citrobacter diversus* under aerobic environment. Applied Microbiology and Biotechnology. 2001;**55**:90-94. DOI: 10.1007/s002530000363

[31] Joo H, Hirai M, Shoda M. Characteristics of ammonium removal by heterotrophic nitrification-aerobic denitrification by *Alcaligenes faecalis* No. 4. Journal of Bioscience and Bioengineering. 2005;**100**:184-191. DOI: 10.1263/jbb.100.184

[32] Liang S, Zhao M, Lu L, Wang C, Zhao L, Liu W. Isolation and characteristic of an aerobic denitrifier with high nitrogen removal efficiency. African Journal of Biotechnology. 2011;**10**:10648-10656. DOI: 10.5897/AJB11.569

[33] Patureau D, Godon J-J, Dabert P, Bouchez T, Bernet N, Delgenes JP, et al. *Microvirgula aerodenitrificans* gen. nov., sp. nov., a new gram-negative bacterium exhibiting co-respiration of oxygen and nitrogen oxides up to oxygen-saturated conditions. International Journal of Systematic Bacteriology. 1998;**48**:775-782. DOI: 10.1099/00207713-48-3-775

[34] Takenaka S, Zhou Q, Kuntiya A, Seesuriyachan P, Murakami S, Aoki K. Isolation and characterization of thermotolerant bacterium utilizing ammonium and nitrate ions under aerobic conditions. Biotechnology Letters. 2007;**29**:385-390. DOI: 10.1007/s10529-006-9255-8

[35] Chen J, Strous M. Denitrification and aerobic respiration, hybrid electron transport chains and co-evolution. Biochimica et Biophysica Acta. 2013;**1827**:136-144. DOI: 10.1016/j.bbabio.2012.10.002

[36] Li Z, Wang S, Zhang W, Miao L, Cao T, Peng Y. Nitrogen removal from medium-age landfill leachate via post-denitrification driven by PHAs and glycogen in a single sequencing batch reactor. Bioresource Technology. 2014;**169**:773-777. DOI: 10.1016/j.biortech.2014.06.076

[37] Mino T, Van Loosdrecht MCM, Heijnen JJ. Microbiology and biochemistry of the enhanced biological phosphate removal process. Water Research. 1998;**32**:3193-3207. DOI: 10.1016/s0043-1354(98)00129-8

[38] Lee H, Min YM, Park CH, Park YH. Automatic control and remote monitoring systems for biological nutrient removal on small wastewater treatment plants in Korea. Water Science and Technology. 2004;**50**:199-206. DOI: 10.2166/wst.2004.0377

[39] Puig S, Corominas LI, Vives MT, Balaguer MD, Colprim J. Development and implementation of a real-time control system for nitrogen removal using OUR and ORP as endpoints. Industrial and Engineering Chemistry Research. 2005;**44**:3367-3373. DOI: 10.1021/ie0488851

[40] Vives MT, Balaguer MD, García S, García R, Colprim J. Textile dyeing wastewater treatment in a sequencing batch reactor system. Journal of Environmental Science and Health Part A, Toxic/Hazardous Substances and Environmental Engineering. 2003;**38**:2089-2099. DOI: 10.1081/ESE-120023335

[41] Cassidy DP, Belia E. Nitrogen and phosphorus removal from an abattoir wastewater in a SBR with aerobic granular sludge. Water Research. 2005;**39**:4817-4823. DOI: 10.1016/j.watres.2005.09.025

[42] Alzate Marin JC, Caravelli AH, Zaritzky NE. Nitrification and aerobic denitrification in anoxic–aerobic sequencing batch reactor. Bioresource Technology. 2016;**200**:380-387. DOI: 10.1016/j.biortech.2015.10.024

[43] Al-Ahmady KK. Analysis of oxygen transfer performance on sub-surface aeration systems. International Journal of Environmental Research and Public Health. 2006;**3**:301-308. DOI: 10.3390/ijerph2006030037

[44] Lobo CC, Bertola NC, Contreras EM. Stoichiometry and kinetic of the aerobic oxidation of phenolic compounds by activated sludge. Bioresource Technology. 2013;**136**:58-65. DOI: 10.1016/j.biortech.2013.02.079

[45] Cui Y, Wang S, Li J. On-line monitoring for phosphorus removal process and bacterial community in sequencing batch reactor. Chinese Journal of Chemical Engineering. 2009;**17**:484-492. DOI: 10.1016/S1004-9541(08)60235-9

[46] Serafim LS, Lemos PC, Levantesi C, Tandoi V, Santos H, Reis MAM. Methods for detection and visualization of intracellular polymers stored by polyphosphate-accumulating microorganisms. Journal of Microbiological Methods. 2002;**51**:1-18. DOI: 10.1016/S0167-7012(02)00056-8

[47] Jenkins D, Richard MG, Daigger GT. Manual on the Causes and Control of Activated Sludge Bulking and Foaming. 2nd ed. USA: Lewis Publishers Inc., Chelsea, MI; 1993

[48] Third KA, Burnett N, Cord-Ruwisch R. Simultaneous nitrification and denitrification using stored substrate (PHB) as the electron donor in an SBR. Biotechnology and Bioengineering. 2003;**83**:706-720. DOI: 10.1002/bit.10708

[49] Beun JJ, Dircks K, Van Loosdrecht MCM, Heijnen JJ. Poly-β-hydroxybutyrate metabolism in dynamically fed mixed microbial cultures. Water Research. 2002;**36**:1167-1180. DOI: 10.1016/S0043-1354(01)00317-7

[50] Gerardi MH. Oxidation-Reduction Potential and Wastewater Treatment. New England Interstate Water Pollution Control Commission. Publication and Resources, Interstate Water Report; 2007

[51] Muszyński A, Łebkowska M, Tabernacka A, Miłobędzka A. From macro to lab-scale: Changes in bacterial community led to deterioration of EBPR in lab reactor. Central European Journal of Biology. 2013;**8**:130-142. DOI: 10.2478/s11535-013-0116-2

[52] WEF, Water Environment Federation. Enhanced biological phosphorus removal. In: Operation of Nutrient Removal Facilities. WEF Manual of Practice No. 37; Alexandria, USA. 2013. pp. 151-208

[53] Kujawa K, Klapwijk B. A method to estimate denitrification potential for predenitrification systems using NUR batch test. Water Research. 1999;**33**:2291-2300. DOI: 10.1016/S0043-1354(98)00459-X

[54] Oehmen A, Lemos PC, Carvalho G, Yuan Z, Keller J, Blackall LL, et al. Advances in enhanced biological phosphorus removal: From micro to macro scale. Water Research. 2007;**41**:2271-2300. DOI: 10.1016/j.watres.2007.02.030

[55] Coats ER, Mockos A, Loge FJ. Post-anoxic denitrification driven by PHA and glycogen within enhanced biological phosphorus removal. Bioresource Technology. 2011;**102**:1019-1027. DOI: 10.1016/j.biortech.2010.09.104

[56] Vocks M, Adam C, Lesjean B, Gnirss R, Kraume M. Enhanced postdenitrification without addition of an external carbon source in membrane bioreactors. Water Research. 2005;**39**:3360-3368. DOI: 10.1016/j.watres.2005.05.049

[57] Oh J, Silverstein J. Oxygen inhibition of activated sludge denitrification. Water Research. 1999;**33**:1925-1937. DOI: 10.1016/S0043-1354(98)00365-0